Introduction to the Philosophy of Science

Cutting Nature at Its Seams

Robert Klee
Ithaca College

New York Oxford
OXFORD UNIVERSITY PRESS
1997

Oxford University Press

Oxford New York
Athens Auckland Bangkok Bogotá Bombay Buenos Aires
Calcutta Cape Town Dar es Salaam Delhi Florence Hong Kong
Istanbul Karachi Kuala Lumpur Madras Madrid Melbourne
Mexico City Nairobi Paris Singapore Taipei Tokyo Toronto

and associated companies in
Berlin Ibadan

Published by Oxford University Press, Inc.
198 Madison Avenue, New York, New York 10016

Library of Congress Cataloging-in-Publication Data
Klee, Robert, 1952–
Introduction to the philosophy of science : cutting nature at its seams / Robert Klee.
p. cm.
Includes bibliographical references and index.
ISBN 0-19-510610-5 (clothbound) — ISBN 0-19-510611-3 (paper)
1. Science—Philosophy. 2. Immunology. I. Title.
Q175.K548 1996
501—dc20 96-18851

Printing (last digit): 1 3 5 7 9 8 6 4 2

Printed in the United States of America
on acid-free paper

For Toner M. Overley, M.D.

Contents

Preface

Many students of philosophy of science, as well as many teachers of it, have complained in my presence of the scarcity of philosophy of science texts that are at one and the same time thorough enough not to be superficial, yet written to be accessible to undergraduate readers. I have written this text in the hope of having done something to remove that scarcity. Along the way I have also tried to make what contribution I could to the growing movement within philosophy of science away from the exclusive use of the physical sciences as a source domain for illustrations and examples. Most undergraduate level philosophy of science texts, let us face it, are really philosophy of physics texts. Philosophy of physics is a wonderful and interesting subject matter, but physics is not the whole of science, nor should its philosophy be the whole of philosophy of science. In this text I have made immunology the central illustrative domain of scientific inquiry. Cases and examples are not taken exclusively from immunological science, but they form the overwhelming majority of them. I believe that this focus on one of the premier life sciences of our time has produced a happier and more fruitful result with respect to making the philosophy of science interesting and accessible to a larger audience of undergraduates. I am aware that immunology is not especially familiar to many philosophers, but one advantage of using that domain of inquiry is that all its fundamental concepts, principles, and specialized terms are readily definable, and in many cases even picturable in a way it is notorious that many phenomena of physics are not. Instructors, I believe, will find in the end that it repays their efforts to learn what immunology is contained in this text, for the conceptual richness of immune phenomena in my view outstrips standard physics. Every major issue central to contemporary philosophy of science, from reduction to incommensurability, has a clear illustrative case within immunology, a case it is easier for students to learn and appreciate, than when they are forced into the often arcane mathematics required to get a case from quantum physics across to them.

This text presents a thorough survey of contemporary philosophy of science in the sense that it covers both the positivist model of science and the currently popular alternatives to the positivist model that flow from Thomas Kuhn's watershed work. Many current texts cover one or the other of these models of science rather well but condemn the remaining one to a superficial treatment or to no treatment at all. Several aspects of the text call for special comment. Chapter 4 on holistic models of science presents the correct interpretation of something widely used but even more widely misunderstood, the Quine-Duhem Thesis and the underdetermination of theory by observational data. Getting Quine's brand of holism right is a delicate task, and chapter 4 does so while at the same time insisting on the correctness of realism as a philosophy of science. Chapter 5 on reductionism and antireductionism contains a discussion of the concept of supervenience, a concept of growing importance and use within the philosophy of science, which most texts simply ignore. Chapter 6 on scientific explanation presents two recent models of scientific explanation that I have not seen covered in any other survey texts: Paul Humphreys' aleatory model and Philip Kitcher's unification model. Chapter 7 presents the original Kuhnian model of science with sanity and sensibleness. I do not read Kuhnian philosophy of science as an invitation to participate in wild-eyed intellectual revolution, as it is so often assumed to be by other writers. Kuhnian philosophy of science is not and cannot be all things to all thinkers, and if chapter 7 makes a contribution to rectifying the gross distortions of Kuhn's work that are far too frequent throughout the current intellectual world, that alone would make the text worthwhile. Chapters 8 and 9 cover the new wave of social constructivist philosophy of science in first its nonfeminist forms and then in its feminist versions. While I do not agree with much of what the social constructivists have to say, social constructivism is here to stay, and any survey text in philosophy of science that seeks even minimal adequacy ought to contain a treatment of it at least as thorough as the one contained in this text.

Throughout the text I make no attempt to hide from the reader my sympathies for scientific realism. I am frankly skeptical of any philosopher of science who claims that it is possible to write an effective philosophy of science text while remaining absolutely neutral on the central issue of realism and antirealism. Nothing of benefit to students is accomplished by trying to maintain such a false pose of neutrality about something it is so obviously absurd to be neutral about. It is possible to present fairly all sides to an issue while taking a position oneself. If this were not possible then philosophy per se would not be possible.

To help students in handling the inevitable glut of technical terms necessary to do useful philosophy of science I have placed a glossary immediately following the text. The first time a term or phrase defined in the glossary is used within the body of the text it is highlighted in boldface print.

The illustrations that appear in the text were constructed by the author using CorelDRAW 4.0.

Acknowledgements

I owe thanks to a number of individuals. My colleague Stephen Schwartz read parts of an early draft of this book and was instrumental in encouraging me to pursue it to completion. Rick Kaufman provided valuable help as chair of the Ithaca College Department of Philosophy and Religion by granting me reassigned time to work on the manuscript. I have learned much from philosophical discussions with Steve and Rick over the years, and their philosophical influence on me has made this a far better book than it would have been otherwise. Thanks go also to Dean Howard Erlich and Provost Thomas Longin for their support of my work on this text, especially to Provost Longin for the award of an Ithaca College Provost's Summer Research Grant in 1994. Philip Kitcher read the entire manuscript and provided perceptive criticism and advice that improved the book significantly. His encouragement was also very helpful and appreciated. The comments of several anonymous manuscript readers for OUP were much appreciated. To those of my teachers who most influenced my general approach to philosophy, Jaegwon Kim, Larry Sklar, and Tim McCarthy, I owe the thanks of a grateful student. My current and past philosophical colleagues at Ithaca College, Richard Creel, Carol Kates, Linda Finlay, Eric Lerner, Michael McKenna, and Aaron Ridley have each influenced in their own way the philosophy of science contained in the pages of this book. My editor Robert Miller has been a font of wise counsel and support throughout the whole project. My parents Howard Klee and Marie Klee were willing to encourage and support a son who wanted to become a professional philosopher. For that, and for so much else besides, they are deserving of my deepest gratitude and love. Many thanks go to Donald Grinols for his wisdom and support through stressful times. And last, but never least, I thank Toner Overley, without whose wise, just, and empathetic medical intervention in my life this book would simply never have been written at all.

R. K.
Ithaca, New York
May 1996

*φύσις κρύπτεσθαι φιλει.**

Heraclitus, fragment 123

*The real constitution of things is accustomed to hide itself.
—trans. G.S. Kirk

Introduction

In 1956 in a Massachusetts hospital a man 51 years old was released and sent home to die. A large cancerous tumor had been removed from him, but a number of other malignant tumors had been found—all inoperable. The surgeons had sewed him up in dejected resignation and his case had been filed away. Twelve years later, incredibly, the same man, now 63 years old, showed up in the emergency room of the same hospital with an inflamed gallbladder. Some doctors might have concluded that the original diagnosis 12 years earlier had been in error and think no more of it, but a young surgical resident at the hospital named Steven Rosenberg was not like some other doctors. Rosenberg made a determined search of hospital records, even going so far as to get a current hospital pathologist to pull the original tissue slides of the patient's removed tumor out of storage and reexamine them. The slides showed that an aggressively malignant tumor had been removed from the man twelve years earlier (so it was a sure bet that the inoperable ones had been of the same aggressively malignant type). During the subsequent operation to remove the patient's gallbladder Rosenberg did a bit of exploring in the man's abdomen to see if the inoperable tumors from twelve years earlier had stopped growing. The man had no tumors at all in the places his record from twelve years earlier located them.

To Rosenberg, the man whose gallbladder he had removed presented an absorbing mystery. How had a patient with multiple inoperable cancerous tumors survived for twelve years in the apparent absence of any therapy whatsoever? Such "spontaneous remissions" were not unknown in medicine—they have long been a central obsession of trendy occultists and other kinds of miracle mongers—but Rosenberg wanted to know the detailed *why* of it. He wanted an explanation of it in perfectly natural terms. He wanted to know how the everyday physiological processes of the human body—in this case, the immune system of the human body—could produce such a remission. Where less curious persons might have shrugged their shoulders in amazement and thought no more about it, Rosenberg instead proceeded on the supposition that there had to be some structural physiological basis behind the patient's remission and survival for twelve years, a structural physiological basis that was consistent with the otherwise ordinary causal operations of the human body. Rosenberg wanted to know the causal details of that structural physiological basis. If he could find out those details, especially if they were quantitative details, then the possibility opened up of being able to manipulate the physiology of cancer patients so as to destroy their cancer.

It would take Rosenberg a number of years to come up with a detailed account of that basis, but come up with it he did. Not only did he find an explanation in perfectly natural terms for the original patient's spontaneous remission, but using the theoretical account of immunological processes associated with that explanation he was able to design a complicated therapy that involves artificially growing cancer-killing immune cells outside the body. These cancer-killing immune cells, called LAK (lymphokine activated killer) cells, are injected back into the patient's body in an attempt to intervene and manipulate the patient's immune response to the cancer. Rosenberg and his colleagues have had a considerable degree of success at producing remissions with this therapy but only for specific kinds of cancer—particularly, kidney cancer and skin cancer. Apparently, knowledge of further structural detail is needed in order to be able to design LAK cells that are effective in other kinds of solid-tumor cancers.

In 1964, a few years before Rosenberg's startling encounter with "The Man Who Came Back from Being Certain To Die Soon," two radio astronomers named Arno Penzias and Robert Wilson were engaged in calibrating a large radio antenna receiver located in a New Jersey field. They were preparing to measure the radio noise level of our galaxy, the Milky Way, at high galactic latitudes. It was known that the Milky Way produces random radio noise—radio static—mostly due to ionized hydrogen gas. The intensity of this radio static peaks in the plane of the galaxy where the majority of the galactic matter resides. Penzias and Wilson wanted to measure the radio noise away from the central plane of the galaxy. The intensity of the radio noise away from the plane of the galaxy was expected to be rather faint, and Penzias and Wilson worried that it would not be much more intense than the random electrical noise produced by the antenna's own receiver circuitry. Accordingly, they devised a method for measuring the amount of random electrical noise produced by the receiver circuitry so that the noise generated by their own processing circuitry could then be "canceled out" during processing of the galactic radio noise. After applying their method they were surprised to find that a faint radio noise still remained, apparently coming from all directions in the sky at once. The leftover noise was extremely faint and centered in the microwave range in terms of its wavelength. It was uniform across the sky to a smoothness of about 1 part in 10,000. In no part of the sky was it more or less intense to a degree in excess of 0.0001. Radio intensity can be measured in terms of radiation temperature. The radiation temperature of the puzzling leftover radio noise was a mere 3.5 degrees above absolute zero. What was this amazingly uniform and extremely faint radio noise? Was it genuinely of cosmic origin, or was it an artifact of their equipment or local observing environment? It was possible that it was an artifact of the receiving equipment. Some pigeons had recently nested in the antenna and were known to have left a bit of a mess. But after the pigeon droppings had been cleaned away, the leftover radio noise was still there. It was also possible that the leftover noise was of human origin. But what humanly produced electrical broadcasting would be uniform across the sky to 1 part in 10,000? Human radio broadcasting is directional in a way the leftover noise was not (in 1964 there was not yet the traffic jam of artificial satellites in earth orbit which we have now).

A group of physicists at Princeton University realized that the leftover radio noise was consistent with a **theory** about the origin of the chemical elements that had been proposed a couple of decades earlier and then mostly forgotten. For there to be the abundances of the heavier chemical elements that we have in the universe today, there would have to have been a tremendous amount of hot radiation in the very early universe. Without such hot radiation, all the hydrogen in the very early universe would have been "cooked" very quickly into heavier elements. Only if there had been enough hot radiation around to rip apart the nuclei of heavier elements as soon as they were formed would there be as much hydrogen still around as we now observe there to be. The hot radiation would have to have been so hot and so intense that a remnant of it would still be detectable today, 15 billion years later. Calculations made at the time—since revised—estimated that the hot radiation would have cooled in that 15 billion years from about 100,000,000,000 degrees above absolute zero to around 5 degrees above absolute zero (we now know the temperature of the radiation remnant Penzias and Wilson discovered to be 2.735 degrees above absolute zero). The two radio astronomers had detected the faint remnant of the very hot radiation produced in the first few seconds of the universe's history. In the popular media it was described as the remnant of the primordial "fireball" from the "big bang" beginning of the universe. Among astrophysicists the radiation was named the cosmic microwave background or CMB.

Both the preceding cases illustrate the important point that science involves a practical engagement with the natural world. This point is often lost sight of when science is portrayed by both its supporters and critics as a largely intellectual enterprise, as a game consisting mostly of free-form conceptual speculation. The process through which science produces knowledge is more constrained than that, and it is messier in the bargain. Rosenberg stumbled down a number of blind alleys for years before he was able to find those structural details to the human immune system that allowed him to grow LAK cells. The CMB was for a number of years considered quite problematic because its explanation involved something impossible to investigate under artificial laboratory conditions: the big bang beginning of the universe. A few skeptics about the standard explanation of the CMB still remain.

In both of these cases we see the nuts and bolts of everyday science at work. A surprising observational datum is noted, and a detailed investigation of the facts surrounding it produces even more puzzles that demand further explanations. It soon becomes obvious to the scientific investigators that most of those further explanations involve regions of the universe not directly observable to human beings. The separation of the natural universe into regions that are observable to humans and regions that are nonobservable or theoretical to humans is nearly universal among scientists. The presumption that the terms of a scientific theory can be classified either as **observational terms** or as **theoretical terms** is, however, a highly controversial presumption, as well as a central one, within the philosophy of science, and a good deal of the first half of this text will contain a thorough investigation of whether a coherent distinction between the observational and the theoretical can be sustained.

Rosenberg wanted a **scientific explanation** of what was observationally unquestionable—the spontaneous remission of his gallbladder patient's previous cancer. Many philosophers consider explanation to be the main business of science. Any philosophy of science worth the time to study must contain some account or other of scientific explanation—what it requires and how it differs from other kinds of explanation. Accordingly, we shall spend a significant amount of time in the text below investigating some of the more interesting and important models of scientific explanation.

In the situation in which Steven Rosenberg found himself, if anything, there were too many competing possible theoretical explanations. Within the philosophy of science this interesting problem is called the **underdetermination of theory**. It is a central issue of concern in contemporary philosophy of science, and, accordingly, its presence can be detected throughout the body of this text. Some philosophers of science argue that the underdetermination of theory by observational data implies that no theoretical explanation in science can be said to be correct, for there are always possible alternative explanations inconsistent with the given explanation. An important goal in philosophy of science, and therefore in this text, will be to see if this is a true or false charge against science. For now, it might be instructive to consider what Steven Rosenberg did in the face of the underdetermination of theory. His task was to pare down the number of competing possible theoretical explanations for his patient's spontaneous remission in a principled and rationally defensible way—in an evidence-based way and not in an arbitrary way. He used the finely detailed causal structure in the human immune system to do so; and, in that way, he let the independent causal structure of the world push back against his own hypothesizing to do the paring down. Thus, Rosenberg's general procedure was evidence driven in a way that, say, theological speculation and political rhetoric are not. The latter two practices are not forms of knowledge gathering primarily driven forward and constrained by independent causal evidence. The constraining impact of independent causal evidence makes science over the long run self-correcting toward greater accuracy or truth, at least it does according to one school of thought within the philosophy of science.

That scientific explanations not only seek the truth about phenomena but actually achieve it in a number of cases is a controversial claim these days, but one which this text will defend. Those who claim that science does deliver and in fact has delivered true explanations of inquiry-independent phenomena on occasion are called scientific realists. This text is an extended defense of scientific **realism** against the opposing position of scientific **antirealism**. Recently, the dispute between realists and antirealists has heated up within philosophy of science as well as other areas of philosophy. Certain antirealists advocate a model of science called **social constructivism** that has recently become quite popular in various segments of the intellectual community. Social constructivists argue that the substantive results of science are invented or constructed through organized social behavior, rather than discovered. An interesting dispute therefore arises about whether science discovers the structure of a preexisting external world of causally independent entities and processes, or whether science invents what we call the physical universe. Much of the latter half of this text contains an extended investigation of whether or not this social constructivist view of science has any merit to it. If it does have merit, then it would seem to follow, for example, that Penzias and Wilson did not discover the CMB; rather, the CMB was an artifact of their own training, equipment, and interactions with peers. The CMB would be a socially constructed phenomenon, one that is dependent on an idiosyncratic collection of accidental social practices, which include but are much larger than the goings-on in a radio antenna in a New Jersey field.

Many philosophers of science find this social constructivist view of science to be implausible. For these philosophers the social constructivist picture of things cuts against the structural unity of the natural universe. The universe is a cosmos, on this view, a complex of interconnected systems, and human beings are of a piece with the rest of nature. Social constructivist antirealism would seem to cut us off from this fabric or net of nature in a way that smacks suspiciously of **relativism**, the view that what the facts are is relative to changeable human concepts and practices. To critics of social constructivism such relativism has the odor of unjustifiable self-importance, as if the human species is a special exception at the center of the universe, as if we are so central and important that the structure of nature is itself relative to our concepts and practices. To some of these antirelativist philosophers, the sciences ought to form a single web, a hierarchy of domains of inquiry. The weblike interconnection of all the sciences simply mirrors the structural unity of the natural universe that science investigates. This picture of things is called **reductionism** in the philosophy of science, and a major issue we shall discuss in this text is whether those who argue for **antireductionism** are correct that the sciences do not and cannot form such a single network as reductionism requires.

Reductionism is where science turns blatantly metaphysical in the sense of turning **ontological**. One of the reasons science is of such intrinsic interest and controversy is precisely because it purports to tell us what the universe is made of, how what it's made of is structured, and why what happens in it happens. The branch of metaphysics traditionally concerned with "what there is" in the universe is called ontology. This text presents philosophy of science as a field having a profoundly heavy impact on ontology. The reader is advised that not all philosophers of science would take such an approach nor approve of another philosopher of science who does. I accept their challenge. It shall be my task to show the richness, cognitive worth, and philosophical interest of an openly ontological conception of science. In my view, what matters most about science is that it purports to tell us what, for lack of a better phrase, I shall call the actual way things really are. Whether it is plausible to believe that science can achieve such a lofty goal we are about to find out.

1

The Case of Allergic Disease:

From Everyday Observation to Microstructural Explanation

It has not always seemed reasonable to human beings to believe that the natural world is size layered in both directions. In one direction there was never a problem. It was obvious even in ancient times that there are objects larger than human bodies, for the evidence of the human senses testified to the enormous size of the earth, for example. But that there are also objects smaller than the smallest objects visible with the naked eye, smaller even than dust motes, was a hypothesis that, for almost 2,000 years from the time of the earliest scientists of classical Greece, remained a purely speculative belief, supported only by "metaphysical" arguments of doubtful plausibility.

Fortunately, things do change. During the past 300 years scientists have succeeded in uncovering more and more fine-grained structure in nature on a smaller and smaller scale of size. Using artificial devices specially constructed to interact with this minutely tiny level of reality, practitioners in various fields of inquiry have claimed to discover the causal mechanisms by which larger scale macroscopic phenomena are produced out of smaller scale microscopic phenomena. This sounds like a simple matter when presented as glibly as I've just now described it; but a large number of problems—some **methodological**, others **epistemological**, and even a few that are ontological—fester at the heart of this otherwise happy picture of things. Notoriously thorny conceptual problems arise concerning the nature of the relation between the micro and the macro. To give you a taste for just how thorny, consider some of the problems that have occupied philosophers of science in recent times. Is the macro reducible to the micro—if so, how, if not, why not? Does the micro explain the macro and never the reverse—if yes, why the asymmetry; if no, why does macroexplanation seem incoherent? Is the relation between them constructional (part/whole) or even definitional? Is causation or some other kind of determination the best account of the determinative relation between the micro and the macro? Are some macrophenomena impossible to find below a certain level of physical microstructure—if so, are such macrophenomena debarred from being given a scientific characterization?

This text will examine the above problems and many more besides. It serves as a general introduction to philosophy of science for two reasons. First, the assortment of conceptual problems that the text focuses on are problems found universally throughout the sciences—be they physical, psychological, social, or so-called "human" sciences. Second, the illustrative sample science around which the text is organized, contemporary immunology, is not only a "hot" new field that will move into a prominent position within philosophy of science in the near future, but it is also an entirely typical science whose history presents a standard case of how knowledge and experimentation evolve within a given domain of scientific inquiry. The lessons learned from immunological science are easily generalizable to other sciences.

1.1 Experimentation and Research Science

The main feature that distinguishes science from other modes of knowledge gathering is the use it makes of observational experimentation. Scientific procedure is never wholly passive nor purely contemplative. In some domains such as theoretical physics practitioners engage in *Gedankenexperimenten*—'thought-experiments' of a highly conceptual or philosophical nature. But even these thought-experiments are constrained in their invention by hard data from observational experimentation; they are not just free-form fantasies, but rather they are attempts to test our understanding of critically important theoretical concepts by placing those concepts under artificial conceptual stress. This is a version of what happens in ordinary cases of physical experimentation. In the more typical case, practitioners of a science must busy themselves with the chore of manipulating the causally relevant structure in the system under study in such a way as to place certain theoretically important aspects of that causal structure under stress. Then the practitioners observe the system to note what happens. In science you find out what the facts are, what the story is, by doing, not by merely thinking; or rather, to be more exact, you find out what the story is by a skillful combination of doing and thinking.

The aspect of observational experimentation that is of greatest importance in science is not the degree to which observational experience confirms the practitioner's expectations, but rather the way in which it upsets those expectations.

It is the occurrence of what the practitioner did not expect to happen that fuels the forward movement of theoretical knowledge within a given field of science. Some philosophers of science have suggested that this feature of science is responsible for its "intellectual honesty," for how and why science differs from ideology. The philosopher of science Karl Popper argued that in genuine science the practitioner is able to specify in advance of testing a theory what disconfirming evidence for the theory—a negative outcome—would look like. Popper claimed ideologues were unable, as well as unwilling, to do the same for their pet ideologies. Most theories are false. A mature scientist recognizes that fact, argued Popper, and is accordingly skittish about strong personal commitment to a given theory (or at least this ought to be the attitude taken). In fact, some philosophers of science bemoan something they call **the pessimistic induction**, which they claim is supported by the history of science. According to one version of the pessimistic induction: on average, any scientific theory, no matter how seemingly secure, will be shown to be technically false within a sufficiently long time interval (say, 150 years). The history of immunology provides a useful illustration of many of the preceding points. It illustrates how critically important the unexpected oddities of observational experimentation are to science, how prepared scientists are to find the negative experimental outcome, and how tentatively truth is ascribable to scientific theories.

1.1.1 Observable Mysteries

In 1893 Emil von Behring, then nine years away from winning the Nobel Prize in medicine, was busy investigating the properties of diphtheria toxin, the biochemical by-product of diphtheria bacteria that is responsible for the disease of the same name. This toxin acts as a kind of poison to normal tissues. A few years earlier von Behring and his colleague Shibasaburo Kitasato had performed an experiment that showed that immunity to diphtheria was due to antitoxin elements, "antibodies," in the blood. What von Behring did not expect to find in his studies on diphtheria toxin—but to his surprise did find—was this: some animals given a *second* dose of toxin too small to injure an animal when given as a *first* dose, nevertheless had drastically exaggerated harmful responses to the tiny second dose. In some cases the response to the puny second dose was so overwhelming as to cause death. Von Behring coined the term 'hypersensitivity' (*Überempfindlichkeit*, in the German) to describe this exaggerated reaction to a small second dose of diphtheria toxin. This experimental finding was so odd relative to the rest of immunological science at the time that it was essentially ignored for about ten years (von Behring's finding was what the philosopher of science Thomas Kuhn calls an anomaly, a result that doesn't fit in with what else we know, and many anomalies in science are simply noted and then ignored when they first appear on the scene—see chapter 7).

In 1898, Charles Richet and Jules Hericourt reported the same finding, this time with a toxin derived from poisonous eels. It too was noted and then ignored. Then in 1902 Paul Portier and Richet published an experimental result that caught the sustained attention of other immunologists. They reported the same exaggerated response to a second small dose of poison derived from marine invertebrates. What distinguished their report of the same phenomenon von Behring first described nine years earlier was their careful and detailed description of the hypersensitive response as an observable form of cardiovascular shock. Richet and Portier worked in France rather than in Germany, unlike von Behring, and a good deal of political tension and professional animosity existed between those two leading centers of immunological research. The French scientists weren't about to use a term like 'hypersensitivity' invented by a German, so they called the exaggerated response *anaphylaxis* (to highlight its harmful aspects as contrasted with *prophylaxis*, the medical term for 'protection').

During the next decade a host of prominent immunologists systematically investigated the nature of anaphylaxis, both its qualitative and its quantitative aspects. In 1903 Maurice Arthus performed the experiments that would result in the discovery of the phenomenon named after him: The Arthus reaction is a characteristic skin lesion formed by the intradermal injection of certain kinds of proteins. In 1906 Clemens von Pirquet and Bela Schick studied serum sickness, the unfortunate phenomenon whereby a small percentage of persons given standardized diphtheria or tetanus shots, which do not harm a majority of recipients, nevertheless become extremely sick from the shots. They argued that the observational evidence pointed to an immunological cause of serum sickness. To have a convenient way of referring to any medical condition in which otherwise harmless or beneficial substances paradoxically produce illness in certain persons who come into contact with them, von Pirquet and Schick coined the term *allergy* (from the Greek *allos ergos*, altered working). In the same year, Alfred Wolff-Eisner published a textbook on hay fever in which he presented the evidential case for hay fever being a form of hypersensitivity traceable to the immune system. In 1910 Samuel Meltzer made the same kind of case for asthma as a form of immunological hypersensitivity somehow localized in the lung tissues.

Notice in this account of the early days of modern immunology how a surprising observational mystery is first noted, then perhaps ignored for a bit, and eventually set upon with experimental frenzy. Not all observational mysteries are happily resolved in such a way (some are ignored permanently); but in a large number of cases the course a given area of science takes does seem *evidence driven* in a way many other forms of knowledge gathering are not driven by observational evidence. Scientific claims deliberately run a risk: the risk of being shown to be false of the world. Again, some philosophers of science have seen in this at-risk status an important contrast with other forms of human belief such as political ideology, theological doctrines, and so on. But we may ask, toward what does science allow itself to be driven by experimental evidence? To put the question another way: In science, how does it come to pass that the quality of surprise that attaches to an unexpected finding is removed and replaced with a familiar understanding of that finding? Rather than attempting to answer this question with a vague overgeneralization based on something as fallible as commonsense intuition, let us continue the story about the immunological theory of allergic disease and see how the mystery was removed in a real case. As philosophers of science, part of *our* evidence base is the actual history of science.

1.1.2 Physical Microstructure

Removing experimental mystery in science is an exercise in providing a scientific explanation for the puzzling observational data associated with the mystery in question. What makes an explanation a scientific one is a question of such importance and complexity that it shall receive its own chapter (chapter 6). For now we need not worry about the absence of a fully detailed model of scientific explanation; what is currently of interest, for our purposes here, is the preliminary fact-finding, the prior searching out of structural detail, which is a precondition of scientific explanation.

That there are allergies, that they exist as observable states of disease, was established by 1911. But the causation of the state of illness observed in allergies was a matter still quite unresolved. Here we see the first entrance of the "theoretical" into our story. The first two decades of the twentieth century witnessed the growth of a number of theories about allergic illness—about all kinds of immunological phenomena, for that matter. What makes a theory a scientific one is a question of such importance and complexity that it shall have two chapters of its own below (chapters 3 and 4). For now, instead of attempting to formulate a sufficiently abstract and thorough characterization of the beast, let us do two things: First, we will note a few important aspects of theories and clear away certain misconceptions people have about them. After we have finished with that, we will look at some sample theories that were offered as accounts of allergic disease immediately after its discovery.

One important aspect of a scientific theory about a phenomenon is that it seeks to unravel and present in a clear way that phenomenon's *causal* properties and relations. If I presented you with a theory about, say, human anxiety, it would be an empty theory if it did not tell you what causal effects anxiety typically produces in human beings as well as what other states of affairs in turn typically cause anxiety in humans. You would be justified in feeling explanatorily cheated if my theory could not causally relate in an understandable way, or at least in a predictively useful way, states of anxiety with other psychological states of mind.

To call a collection of statements a theory is not to suggest that those statements are especially doubtful or highly conjectural. Many theories are composed of statements so well confirmed and confidently held that only an extremely neurotic person could doubt their truth. The term 'theory' does not mean something like 'a bunch of statements just as likely to be false as to be true', although some pseudoscientists—creationists, for example—seem to write and speak as if they think it does mean that.

Another mistaken notion about theories is this: Some people think that calling a collection of statements a theory is an insult to it, for to such persons the term means that what the statements in question claim hasn't been "proven," and that is supposedly a bad thing, according to these people. Unfortunately, they have an erroneous conception of what it is to give a **proof** of the truth of a statement, and they mistakenly think that all kinds of so-called "proven facts" exist—as contrasted with these supposedly impoverished things called theories—and the existence of these proven facts supposedly implies that their poorer cousins, theories, are nothing more than guessing games. The bogus idea here is that a theory is just a would-be fact, a wanna-be that doesn't measure up to the requirements for being a legitimate piece of human knowledge. Hogwash. A theory is a map of the domain of inquiry. Like more familiar maps, road maps for example, a theory provides a way of proceeding through the structural reality of the domain in question. Depending on one's purposes, the same map can be used in very different ways. And, just like more familiar maps, a necessary element of idealized distortion creeps into a theory. The smaller the idealized distortions, the better the map/theory. Now a map, if it is a good one, hardly qualifies as "unproven." Proof, in a sense, isn't the important issue. What is important is that the map be readable and that it serve our navigational purposes. But to serve our navigational purposes, the map must be relatively accurate. It cannot put a lake where a desert is. In the same way, a scientific theory will eventually fall into disuse unless its accuracy with respect to the phenomena it is about meets a certain minimum standard. In this way the world "pushes back" against our theorizing. So to call a collection of statements a theory is hardly to insult it; rather, it is really a compliment of sorts, for a theory is a much more sophisticated and impressive beast than a mere proven fact. A theory is an attempt to organize the facts—some "proven," some more conjectural—within a domain of inquiry into a structurally coherent system. And understanding how a system works empowers human beings a lot more than understanding a few isolated proven facts. A frequent contrast some critics of science often use is that between theories and so-called "hard data." What those critics fail to realize is that there is no such thing as hard data minus some theory under which they count as hard data. It is a theory that tells the observer that what is being observed is relevant in the right way to what the observer is interested in finding out. Before the rise of Darwinian biology and modern geology, for example, the bones of ancient animals found deep in the soil were not appreciated as being fossils of any kind. They were often not observed as important or relevant because the prevalent theories of the time—usually religious in nature—did not recognize the possibility of evolution from common descent over a vast time scale. Observations become *data* because other beliefs informed by theories already held confer the status of being relevant data upon those observations.

So it was perfectly appropriate that immunologists struggling with the birth of modern allergic medicine should have taken to constructing theories about allergic phenomena. What they sought was a causal model that would explain how the observable symptoms of allergic disease are brought about by the interaction of the microstructural components of the immune system. What made this a daring project at the time was the fact that very little was known about those microstructural components. Such ignorance rendered the theorizing quite risky, and there were bound to be many wild stabs that fell wide of the mark. It wasn't clear, for example, whether the mechanism of allergic disease was due to the action of antibodies (soluble proteins) or immune cells. Some immunologists who supported the antibody side of the dispute speculated that perhaps antibodies came in two general kinds, mostly good ones and a few bad ones, and that allergic symptoms are caused by the bad ones. Other theorists who supported antibodies thought that perhaps allergies were the result of a "precipitate" formed by the interaction of the allergic substance and "ordinary" (good) antibodies. Supporters of the cellular model of allergies pointed to the fact that anaphylactic shock could be produced in experimental animals in the complete absence of detectable blood antibodies. These theorists proposed that allergic symptoms were produced by the interaction of an as yet undiscovered class of immune cells and the allergy-producing substance. One interesting problem that we can now see looking back on the early days of allergic theory is that there was a general reluctance to see allergic phenomena as immunological at all, for there was a widespread presumption that the immune system could only do an organism good, and it would not contain components that would naturally lead to disease in the course of carrying out their normal immunological functions. In this spirit some theorists suggested that allergic states of the immune system were a normal stage in the development of an ordinary immune response, a stage that somehow went out of control in allergic individuals.

Slowly the microstructural facts emerged from out of the wealth of experimentation. A major experiment which conclusively demonstrated that allergic disease was a function of blood-borne antibodies and not immune cells was the passive transfer reaction demonstrated by Carl Prausnitz and Heinz Kustner in 1921. Passive transfer was a general experimental procedure that had been used for a number of years by immunologists. In passive transfer one demonstrates that a given immunological sensitivity must be blood-based in nature because one is able to transfer that sensitivity reaction by injecting blood serum into a second organism never exposed to the sensitizing substance. Kustner was extremely allergic to certain fish. When extracts of such fish were mixed with some of Kustner's blood, however, no reaction was observable. Yet, when a small amount of Kustner's blood was injected into Prausnitz's skin along with the fish extract, an allergic reaction was observable at the site of injection within twenty-four hours. Thus, Kustner's allergy to fish could be passively transferred to Prausnitz; hence, the causal mechanism of allergic disease must be blood-based, and that meant it was antibody-based. The cellular theory began to wither away. It withered away not because it had been conclusively disproven, but because the cost of adjusting the cellular theory of allergy to make it consistent with the passive transfer results was too high: The theory would have to become too complicated, with special adjustments added to explain away passive transfer. This is another feature of science which any philosophy of science must account for: Rejected theories are rarely conclusively disproven, they usually are abandoned on grounds of there being a competitor theory with better explanatory and predictive power.

1.2 Mechanism and Testing

When practitioners score a hit with their theories nature begins to reward them. Experimentation begins to produce generalizable results that fit together into an overall pattern rather than haphazard and isolated findings that increase confusion rather than lessening it. The development of allergic medicine was delayed a bit by World War II. After the war, however, within roughly twenty years, all the essential microstructural features of allergic phenomena had been worked out.

1.2.1 Mast Cell Release

In 1910 Henry Dale and Patrick Laidlaw had shown that anaphylaxis could be produced by the injection of histamine into appropriate tissues. Several following experiments confirmed the important role played by histamine in the induction of allergic responses. Immunologists, however, were not sure how naturally produced histamine entered into the picture. In 1953 James Riley and Geoffrey West cleared away that mystery by showing that histamine was contained preformed inside mast cells. Mast cells are found in the skin, intestines, and most mucous membrane tissues such as those that line the upper respiratory tract (nose, throat, and lungs). An obvious causal mechanism for allergies could be constructed from this result. Somehow, antibodies that interact with allergy-producing substances must do so in such a way as to release the preformed histamine from inside mast cells lining the skin, intestines, and upper respiratory tract. The histamine then produces the typical inflammation of allergy by ordinary chemical means. This theory is basically correct in fundamentals if a bit oversimplified. We now know that mast cells aren't quite this simple: They contain a host of different inflammation-producing biochemicals, histamine being merely one of them.

1.2.2 Immunoglobulin E

Research in allergic immunology for the next two decades focused on the discovery of how mast cells are forced to release their contents into the surrounding tissues. The discovery was made in 1966 by the team of Kimishige and Teruko Ishizaka. A little background is all we need to appreciate the ingenuity of how the Ishizakas discovered the mechanism of mast cell release. Not all antibody molecules are of the same kind. It had been known for many years that antibodies come in a small number of classes or kinds. Perhaps the class most people are familiar with is the gamma or G class of antibodies. Shots of gamma globulin antibodies are given to many people for a variety of medical conditions (rabies, malaria prevention, and so forth.). The word 'globulin' is the physiological term for 'protein', and antibodies qualify as a species of proteins called immunoglobulins (that is, 'immunologically active proteins').

By 1966, four general classes of antibodies had been discovered, and they had been given scientific names based on letters: immunoglobulin A, immunoglobulin D, immunoglobulin G (the most frequent class in the human body), and immunoglobulin M. The Ishizakas produced an antibody that attacked "something" in the allergenic fraction of the blood of a patient extremely hypersensitive to ragweed pollen. In other words, they produced an antibody to another antibody. This is perfectly possible in the mammalian immune system. The Ishizakas wanted to be sure that their antibody attacked the antibody in the allergic patient's blood that was responsible for that patient's allergy to ragweed pollen. They confirmed that it did so by showing that their antibody neutralized (cancelled out) the passive transfer of the patient's ragweed allergy with his serum. That is, they used the Prausnitz-Kustner passive transfer experiment to show that their antibody was specific for the antibody causing the ragweed allergy in the hypersensitive patient: When they mixed their antibodies with the patient's blood and injected the mixture into the skin of an unsensitized organism, no allergic reaction occurred because their antibodies had neutralized the patient's antibodies to ragweed pollen.

It remained for the Ishizakas to purify their antibody so that its class could be identified. Was it a form of immunoglobulin A, D, G, or M? The answer was none of them. The purified antibody would not react with any of the four known classes of antibody. The Ishizakas inferred that their antibody must be from a fifth class of immunoglobulins, a class which was eventually named immunoglobulin E (for 'erythema', the medical term for the redness of inflammation).

The discovery of immunoglobulin E was followed shortly by a number of technological improvements involving the ability to identify and isolate individual molecular and cellular components of the immune response. Eventually the microstructural mechanism of allergic disease was worked out in precise detail in the 1970s and 1980s. Immunoglobulin molecules—antibodies—are complex proteins that fold in three dimensions into roughly Y-shaped molecules. The end with the Y-arms contains the part of the antibody that binds to its target molecule. The opposite end of the antibody contains the portion of it that, in some classes, is capable of binding, not to its target molecule, but to structures on the surfaces of various cells called receptors. Thus, some antibodies can fix one end of themselves to, say, the surfaces of mast cells while their other ends are fixed to their target molecules. All antibodies regardless of which class they belong to have the same general chemical structure. Figure 1.1 contains a schematic representation of a typical G-class antibody. The classes differ chemically in the composition of certain parts of their heavy chains, especially the parts of the heavy chains near the c-terminal of the molecule.

It turns out that immunoglobulin E molecules are manufactured by a class of white blood cells called B cells, which are stimulated to produce E-class antibodies to allergy-producing substances (allergens) after fragments of those substances are presented to them by yet another class of cells called APCs (antigen-presenting cells). This process by which E-class antibody producing B-cells are primed and stimulated to produce antibodies is the sensitization phase of allergic disease, and it is illustrated in Figure 1.2. The E-class antibody molecules produced during the sensitization phase are capable of attaching themselves in pairs to the outer surfaces of mast cells, with their allergen binding ends exposed. On subsequent exposure to the same allergen molecule, say a molecule of rag-

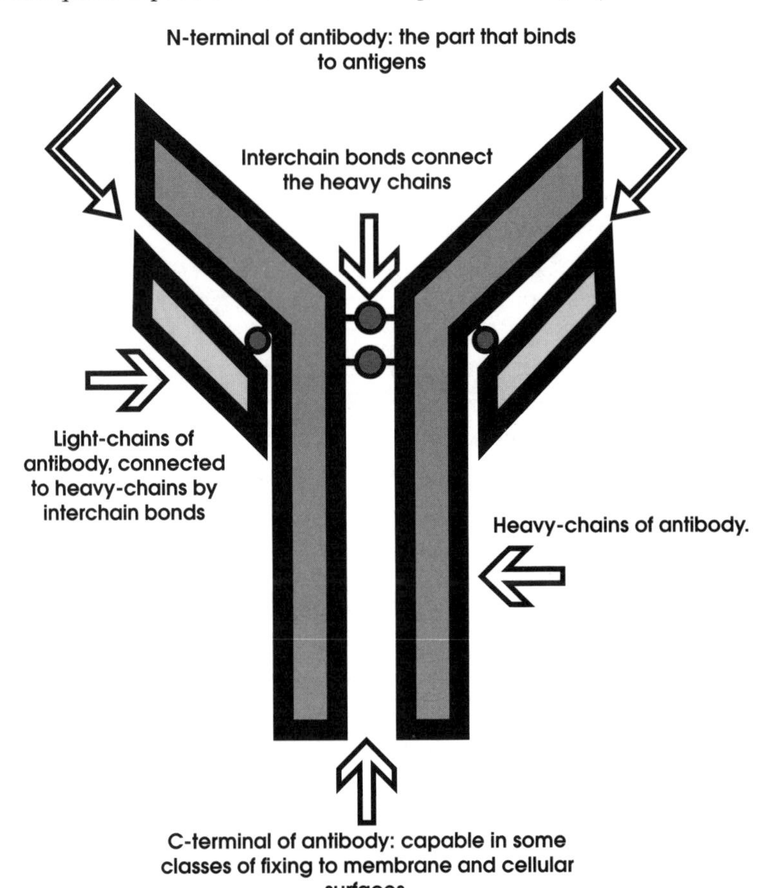

Figure 1.1. The structure of a G-class antibody.

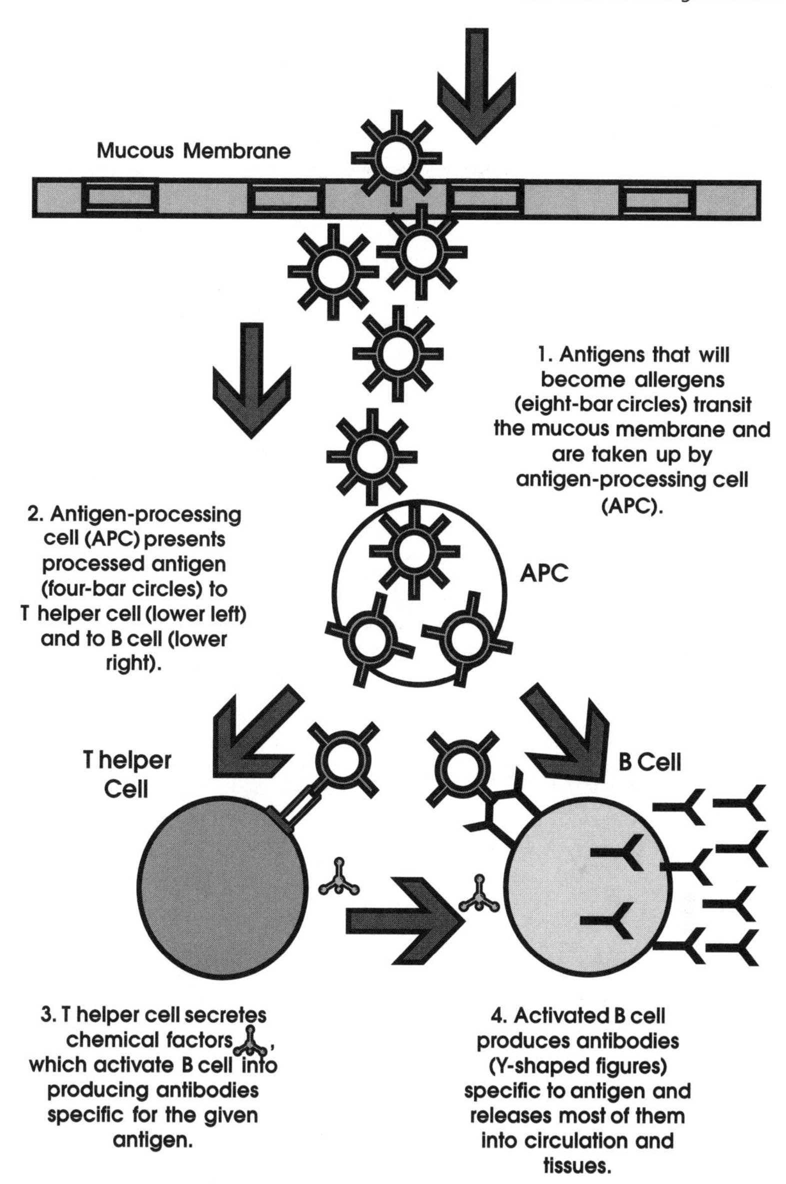

Figure 1.2. The sensitization phase of allergic disease.

weed pollen, the allergen molecule binds one arm each of the two antibodies next to each other, a phenomenon called cross-linked binding. The cross-linking causes a biochemical cascade of events that results in the rupturing of the mast cell's membrane, releasing granules of inflammatory substances like histamine into the local tissues. The result is a local allergic reaction. This inflammatory phase of allergic disease is illustrated in Figure 1.3.

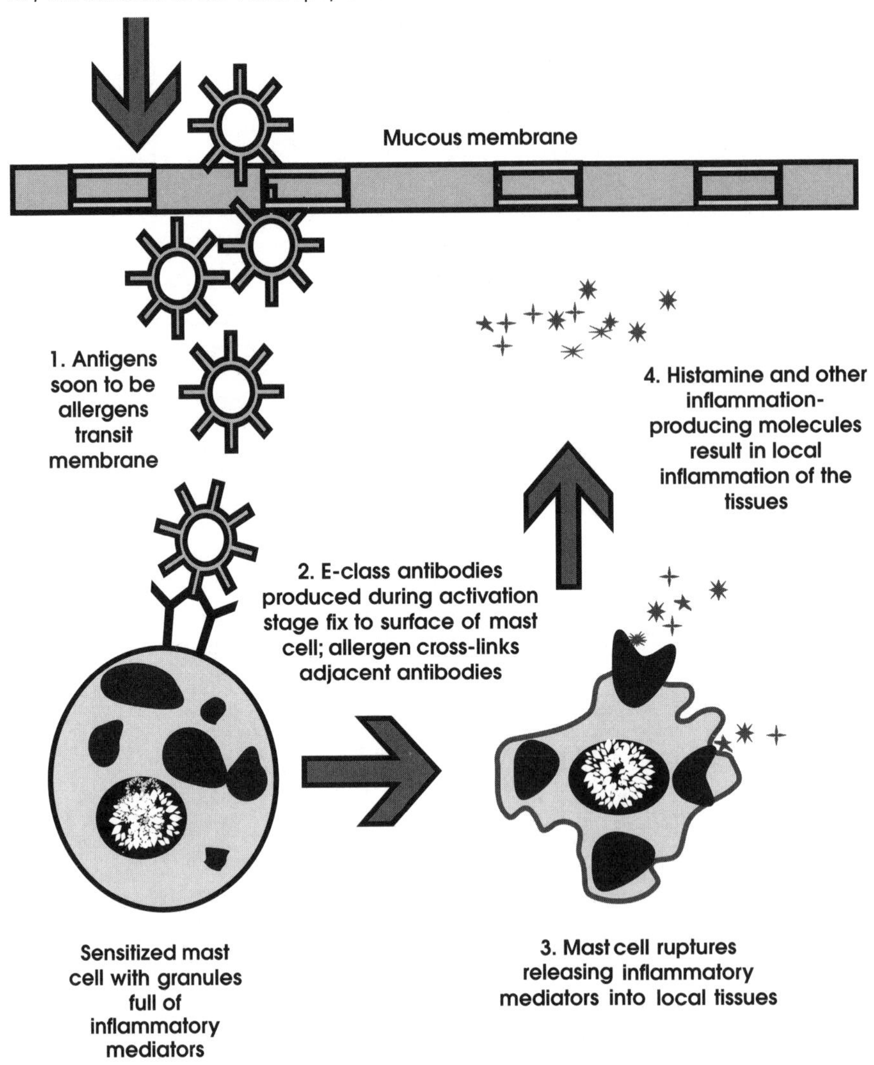

Figure 1.3. The inflammation phase of allergic disease.

1.3 Intervention and Manipulation

A scientific theory seeks to provide a complete account of its subject phenomena. In most cases that means that theorizers must fill in gaps in their theories with the most plausible guesses their training, knowledge, and experience suggest. Very few theories come certified to be 100 percent accurate. This is not a shameful feature of science, but rather a source of its flexibility and honesty. Theories are inherently modifiable in light of new evidence. This prevents the dogmatization and fossilization of scientific knowledge. If any intellectual institution is self-correcting, science would seem to be a model example. As soon as a theory is worked out in sufficient detail to make experimental testing of it possible, practitioners test it ruthlessly. By 'ruthlessly' I mean that scientists are generally skeptical about their theories. They are aware of the element of idealization in all theories and of how theoretical gaps are filled by "best guesswork." This leads to an attitude of healthy skepticism on the part of the practitioner, a skepticism that can be summed up in the belief that the typical scientific theory is sure to be, at a minimum, false in minor detail, and, at worst, a fundamental distortion of the real causal processes it seeks to capture.

To minimize the errors of detail and any fundamental distortions practitioners use their theories to intervene in the phenomena under study; that is, they seek to use the theory to try to manipulate the systems that the theory under test supposedly gives an accurate account of. If their attempts to intervene and manipulate are successful, that success is taken as evidence that the theory used to design the methods of intervention and manipulation is basically correct (and perhaps even correct in most details).

1.3.1 Released Inflammatory Mediator Blockade

The first form of medical treatment for allergies to show any reasonable degree of effectiveness was the use of drugs that counteract the effects of histamine, the antihistamines. These agents are now familiar to most people, for they are common over-the-counter remedies used in everything from cold pills and cough medicines to preparations that combat airsickness and seasickness. Diphenhydramine, familiar to many people under its American brand name of Benadryl, blocks histamine molecules from binding to their target receptors. In some allergic individuals an observable decrease in inflammation results from this blockade. If the role given to histamine in the general theory about allergic disease was in error, if histamine in fact did not participate in the causation of allergic symptoms in the way the theory claims it does, then the successful intervention provided by the ingestion of antihistaminic drugs would be an anomalous phenomenon; it would be an intervention whose success was mysterious. Rather than treating it as a serendipitous accident that antihistamines decrease the symptoms of allergy, surely the more reasonable thing to do is to take it as provisional evidence that the role the theory assigns to histamine in the production of allergic phenomena is basically correct (if not correct in detail). The allergist can successfully intervene through the prescribing of an antihistamine because the theory has the causal part played by histamine correct.

1.3.2 Mast Cell Release Blockade

But intervention at a single point in the causal chain could be a lucky accident (it is not impossible for it to be one). Our confidence in a theory would be bolstered if it could be used to intervene successfully at several different points in the causal chain. Taking antihistamines is a form of treating the symptoms rather than the disease; for once the inflammatory biochemicals have been released by the ruptured mast cells, the damage has been done. We have inflammation on our hands, and the patient is sick. What if we could somehow prevent the mast cells from rupturing in the first place? If we could do that, and then observe a decrease in the observable inflammation, then we would have evidence that the causal role the theory assigns to mast cell rupture is basically correct.

For many years there was no way to intervene in the allergic process so that mast cell rupture could be prevented. But a drug called cromolyn sodium was discovered that appears to coat the outer surfaces of mast cells so that, even if two E-class antibodies are bonded to an allergen molecule fixed to their arms, the mast cell does not rupture and therefore does not release inflammation-producing substances into the local tissues. Cromolyn sodium appears to be quite effective at decreasing observable inflammation in many allergic individuals, and it is used especially in the treatment of allergic asthma. It is an interesting glitch in the story that there is evidence suggesting that cromolyn sodium does not prevent the rupture of mast cells located in the tissues lining the intestines. However, it is clear that it does prevent rupture of mast cells in the respiratory tract and the skin.

Here is yet another successful form of intervention that would be a mysterious anomaly if our current theory about allergic phenomena was completely wrong about the role of mast cell rupture in causing allergic disease.

1.3.3 Immunoglobulin E Blockade

To most people, having allergies is associated with receiving lots of intramuscular shots. Those shots are for the most part not shots of antihistamines or cromolyn sodium (the latter doesn't come in shot form). So what is in those endless shots? You might be surprised at the answer: The shots contain small doses of certain substances to which the person receiving them is allergic. The medical term for such a procedure is hyposensitization, meaning 'to render less sensitive'. It was discovered rather early in the history of allergic medicine that, at least for some allergy-causing substances, giving the allergic person small but increasing amounts of those substances in the form of intramuscular shots eventually desensitizes the person to those substances. They become less allergic over time. It turns out, curiously, that you cannot desensitize in this way to every kind of allergy-producing substance. Animal hair and dander, for instance, cannot be made less allergenic by hyposensitization. The same goes for most foods and fabrics. But many airborne allergens within a certain molecular size range, including all forms of pollen, mold, and dust, can be made less allergenic to the patient by means of hyposensitization.

For many years, precisely why hyposensitizing shots worked was inexplicable. It literally was a mystery, and that mysteriousness fed a general skepticism on the part of many people—doctors as well as laypersons—about whether allergy-shot treatment was all a big myth, a way of pouring money down the drain for worthless shots. Indeed, hyposensitization takes a long time before it begins to cough up results. Many people become frustrated and quit the shots long before they have a chance to begin showing positive results. Thus, the skepticism was and continues to be self-fulfilling. There were and still are many treatment failures, many unsuccessful cases, involving hyposensitization. This part of the causal story is perhaps that part of the theory of allergic disease most subject to controversy, even among practicing allergists. For example, there is a longstanding disagreement among allergists over whether tobacco is an airborne allergen (and therefore potentially subject to successful hyposensitization) or just a membrane irritant (for which hyposensitization is worthless). This is no mere semantic dispute; for entirely different biochemical sequences of events are implicated by each term. In other words, the dispute involves a causal difference within the phenomenon itself.

According to the latest studies, if hyposensitization is done correctly and for a long enough time, positively spectacular results are obtainable. Patients for whom summer was previously a season of interminable suffering find after ten years of hyposensitization shots that they barely notice when the pollen blooms, that they can run a marathon breathing humid and mold-saturated air. Patients who previously panicked when a date announced that he or she was an outdoors person find that after a sufficient number of years of hyposensitization shots they can hike through wild meadows infested with knee-high grasses and weeds for hours without having to suffer an attack of asthma. How is this kind of manipulation of the immune system achieved? I call the result of hyposensitization a manipulation rather than an intervention because the level of interference in allergic disease it is capable of achieving amounts virtually to preventing the allergic reaction in the first place. And this prevention involves using one part of the immune system to shut down another part of it. That is manipulation, not mere intervention. How can injecting an allergic person with the very things they are allergic to result in the eventual elimination of their allergies? The answer is intriguing. Immunoglobulin E antibodies, the ones responsible for allergies, are located in the skin, respiratory tract, and gut—they like to hang out in membranes. But hyposensitization shots are given into a muscle, usually the upper arm muscle. That means that the allergen molecules are picked up first by the blood capillaries and routed into the lymph nodes. In the lymph nodes the allergen molecules are presented not to B cells that make E-class antibodies, but to B cells that make G-class antibodies. Those B cells churn out lots of G-class antibodies to the allergens. Those G-class antibodies migrate eventually into the same membranes where the E-class antibodies hang out. The two classes of immunoglobulins *compete* for binding to the same allergen molecules that come into the membranes and tissues from the outside environment. If there are enough G-class antibodies to the allergens, lots of them because of years of shots, then they bind all the available allergens in the membranes, and the E-class antibodies can find nothing to bind. The causal mechanism which begins with binding of the allergy-producing substances to pairs of E-class antibodies sitting on the outer surfaces of mast cells never gets started; for all the allergen was sucked up by the competing G-class antibodies, which, because of structural differences from E-class antibodies, are not capable of rupturing mast cells. The observable result is the cessation of allergic symptoms altogether. One has treated the disease and not merely its symptoms.

How do they know this is the mechanism? The evidence is presently circumstantial. Patients who have taken hyposensitization shots for many years have elevated blood levels of G-class antibodies in the absence of any other infectious state of their bodies. Tissue secretions from the mucous membranes of such patients are rich in G-class antibodies compared to tissue secretions from the mucous membranes of untreated individuals. Is this proof? Of course not. But it does go together with the other forms of successful intervention and manipulation to produce a high plausibility for the current theory of allergic disease.

1.4. Unification and Explanation

A scientific theory does not live isolated in epistemological space. The universe is a cosmos, not a chaos, and its various levels and parts fit together, albeit roughly, to form an overall fabric. Even chaos theory, of recent interest throughout the sciences, is consistent with this view; for chaos is still a phenomenon explicable by mathematical laws and principles (otherwise it could not be a subject of study at all, and it certainly wouldn't be teachable, which it is). Even in chaotic systems there is a kind of order, just not the sort of strict, Newtonian, deterministic order some people mistakenly thought those systems must exemplify.

One piece of evidence that our current theory of allergic disease was mostly correct would be the finding that it "fits" nicely with what we know about other parts of nature, if it was consistent and congruent with related and collateral areas of scientific knowledge. The term philosophers of science use for this kind of interweaving and fitting together of distinct theories is **unification**. Can the theory of allergic disease be unified with other parts of the scientific scene? The answer is, yes.

1.4.1 Immunoglobulin E, Anthelminthic

A question should have occurred to you by now. Why does the human body even need different classes, different kinds, of antibodies? Why wouldn't nature take the simpler course and just have one kind of antibody? The answer is that, in organisms of our size and complexity, evolution works most smoothly by diversifying biological functions, by dividing the biochemical labors necessary to maintain the health and reproductive fitness of the organism among different organs, tissues, cells, and secretions. Immunoglobulin M, for example, seems to be specially suited to guarding the blood from structurally complex harmful substances. It is the predominate class of antibody coating the membranes of B cells. It rarely migrates from the blood out into the tissues and membranes. Immunoglobulin A, on the other hand, is sometimes called the secretory antibody, because it resides mostly in the mucous membranes. It is the major antibody in saliva, breast milk, and genital fluids. It is structurally adapted to enable it to migrate into these peripheral secretions so that it can serve as the first line of defense against germs attempting to penetrate the body through its various mucous orifices. It is the only class of antibodies with any degree of antiviral activity. Immunoglobulin G, the most numerous class of antibodies, patrols the entire body, but it likes to congregate in the blood and lymph nodes. It is the only class of antibodies that appears effective against bacterial toxins, and it is the only class of antibodies that crosses the placenta to protect the developing fetus. Immunoglobulin D is somewhat the mystery class of antibodies. It is known to coat the surfaces of B cells, but it does not appear to do much else. Current theory speculates that D-class antibodies are involved in the process by which white blood cells, in this case the class of white blood cells called B cells, become activated from their resting states and begin churning out antibodies (it apparently takes an antibody to make an antibody).

These four classes of antibodies were known to the Ishizakas in 1966 (although not all their functions were then known). But the Ishizakas discovered that the antibodies responsible for allergies were not from any of these four classes. Now it would seem bizarre, from an evolutionary standpoint, if the human body contained a class of antibodies whose sole function was to make you sick with allergies. That makes no sense, biologically speaking. There would be no reproductive advantage to an organism's possession of such antibodies. Sickness is debilitating, and any organism with a gratuitous debilitation is surely at a reproductive disadvantage relative to his/her nonsick cohorts. Hence, evolutionary theory demands that there be some positive function that E-class immunoglobulins perform—some function that aids the organism, which increases its reproductive fitness rather than decreases it. Let us be clear what is going on here. The theory of evolution by natural selection is a theory which is logically *independent* of the theory of allergic disease. It is one of those other pieces of the overall fabric of the universe into which allergic phenomena must fit congruently if the universe is to be a cosmos and not a chaos. If allergic phenomena, as portrayed by current immunological theory, can be rendered congruent with the theory of evolution by natural selection, then we have unified these two distinct domains of nature. We have fit the microstructural domain of allergic phenomena into the larger scheme of biological life on the planet earth.

I am happy to report that we can produce the required congruence. Immunoglobulin E, the E-class antibodies, the same ones that cause allergies under certain circumstances, are the only class of antibodies effective against parasitic worms, the kinds of parasitic worms which can infect the human intestinal tract, and which surely did so to many of our ancestors under the harsh conditions of early human life. E-class antibodies evolved as a means to kill off, or at least to reach an equilibrium state with, parasitic worms. A substance that kills worms is called an anthelminthic. E-class antibodies have anthelminthic activity in the human gut—that was their original function. But nature is imperfect, and by a conglomeration of biochemical accidents, these anthelminthic antibodies can cause allergies under the appropriate conditions.

1.4.2 Allergy Genes

The human person is a biological organism. (That claim does not prejudice the issue of what other kinds of being the human person might be: a spiritual being, a social being, and so forth.) As such, the human person is a genetic being. The human body is a genetically based entity, at least this is what the larger realm of biological theory implies. This opens yet another avenue of unification. Can the theory of allergic disease be rendered congruent with contemporary genetics, which is yet another logically independent domain of scientific inquiry.

The answer is, yes. Certain capacities of the immune system, not just the capacity for allergic disease, have been mapped genetically. For example, we know that some members of the class of white blood cells called T cells are capable of recognizing when another cell has become infected with a virus. One of the jobs of such T cells is to destroy virally infected cells. The "recognition" (notice that the use of this psychological term is a metaphor—T cells don't literally *know* anything) process is under genetic control, and the genes that control it are located on chromosomes 7, 14, and especially, 6.

Allergies run in families. It is standard practice in large allergy clinics for the medical staff to file patient charts by first names, not last names. So many members of the same family are likely to be patients at the same allergy clinic that it is more efficient to color code charts by first names. The base rate of allergy among the American population is around 10 percent. Any two nonallergic parents taken at random thus pass a roughly 10-percent risk of allergy to each of their children. Yet the risk of allergy among children of two allergic parents is around 50 percent. With one allergic parent the risk is around 30 percent. These figures clearly suggest a genetic basis to allergic disease; hence, one would expect unifying evidence to be forthcoming from the independent domain of medical genetics.

The immune gene region of chromosome 6 is often referred to as the HLA (human leucocyte antigen) region. To make things more terminologically confusing, this region is also called the MHC (major histocompatibility complex) region. This region is critically important to proper immune function, for a number of human diseases appear to involve genes located in the MHC region on the short arm of chromosome 6, such as rheumatoid arthritis, systemic lupus, and insulin-dependent diabetes.

We can add allergies to the list. If highly purified substances are used, the percentage of responders to the ragweed pollen fragment Ra5 who have the MHC gene HLA-Dw2 is close to 90 percent, far in excess of the gene's base rate among nonallergic individuals. Further, a random survey of patients with positive allergy skin tests to a variety of common substances turned up a frequency for the MHC genes HLA-Dw3 and HLA-B8 more than twice the rate for patients with negative allergy skin tests to the same variety of substances.

1.4.3 Determination and Explanation

What we see happening in the above history of immunology is the coming together of a mosaic, a broader picture, in which the place of allergic disease fits smoothly. It is no longer quite the anomaly it once was at the dawn of the twentieth century. It is understandable substructurally in its own terms, as well as being understandable in terms of its connections to the larger realm of nature. But how could that be, if our theory of allergic disease was in error about the basic causal mechanisms at work in allergic phenomena. How could our intervention and manipulation succeed unless our explanatory scenarios were largely correct, and how could those same explanatory scenarios be largely correct unless the causal structures the theory postulates were largely correct? The answer is, neither could be correct unless the determinative structures postulated by the theory were basically right. *Determination underpins explanation*, and *explanation underwrites intervention and manipulation.* This is the recipe for successful science. The pattern is the same in other areas of scientific inquiry far removed from the level of immunological phenomena. In subatomic physics, for example, the huge cyclotrons, the "atom smashers," simply aren't going to work at a rate above chance if the current quantum theoretical account of subatomic reality were largely in error. As mentioned before, there are always minor errors of minute detail—those are the constant source of challenge for practitioners—but the probability of global error, massive error, is so remote as to be negligible.

The reader should be aware that the preceding sentence would meet with some resistance in many quarters these days. It is currently the intellectual fashion among trendy thinkers and cultural critics to doubt seriously the basic accuracy of mature scientific theories, to pooh-pooh science's claim to tell us "the ways things actually are." These thinkers offer instead a picture of science as "socially constructed knowledge," as a morally questionable institution heavily dependent on changeable social values and hopelessly infected with politicized interests. Science, in this view, is simply one arm of the power elite that controls society; and science is designed to pass off as eternal truths what are really just optional systems for categorizing phenomena that we could reject for systems of more enlightened categories if only we freed ourselves from the authoritarian trance in which science puts us. Science becomes, in the extreme version of this view, a convenient whipping-person—an institution ripe for blaming our worst problems on. Some of these thinkers even suggest that what you want to be true can be arranged, if you just control the reins of the appropriate sciences long enough.

It should be apparent that I do not share this trendy view of science. Frankly, I find it so implausible as to be eccentric, but it is no case of mere eccentricity to many of the cultural critics who support the view. Their views are in some cases thoughtfully argued and therefore they cannot be ignored. Accordingly, we will devote chapters 8 and 9 to a consideration of them.

The tradition in philosophy of science, at least in this century, has conceived the proper unit of analysis to be the scientific theory. I have certainly done so in this chapter. Theories are what the philosopher of science studies, in this view of the matter, and all other issues—explanation, lawlikeness, reduction, and so forth—are to be formulated in the light of theories. In recent years, this approach has come under attack. Some philosophers of science now argue that theories are not the best unit of analysis within the philosophy of science. Their arguments are interesting; and, if sound, their arguments ought to be heeded. Perhaps some other aspect of science besides its theories will offer more fruitful and insightful inroads to understanding science. We shall evaluate these recent alternatives as they crop up in what follows. In the meantime, we will begin with the one model of science—theory-based to a fault, some would claim—which dominated philosophy of science for nearly fifty years until the 1970s: the **positivist model** of science as it was worked out in the positivist conception of scientific theories.

Further Readings

A fundamental reference work in contemporary immunology is *Basic & Clinical Immunology*, 6th ed., edited by D. Stites, J. Stobo, and V. Wells, Norwalk, Conn., Appleton & Lange, 1987. Those who wish for more of a textbook-style approach to immunological science should consult I. Roitt, J. Brostoff, and D. Male, *Immunology* 3rd. ed., London, Mosby, 1993. For a thorough history of immunology see A. Silverstein, *A History of Immunology*, San Diego, Academic Press, 1989. A collection of articles appearing originally in *Scientific American* that cover a host of central topics in immunology is *Immunology: Recognition and Response*, edited by W. Paul, New York, Freeman, 1991.

2

The Positivist Model of Scientific Theories

In the previous chapter we examined a few of the popular misconceptions about scientific theories. Our remarks there were mostly negative: They indicated that scientific theories do not have certain features that many people think they have. Theories are not somehow especially "iffy" (as opposed to alleged "proven facts"), and to call a body of assertions a theory is not to suggest that the assertions are just as likely to be false as to be true. But it would be helpful if we could find some positive characterization of theories. What are they? Are they mental objects? Abstract objects? Are they not objects at all, but rather some sort of linguistic representation of external reality?

The positivists held that a scientific theory is a linguistic representation, not of external reality (which they were not sure was a notion that even made sense), but of actual and possible human experiences. Their idea was that a scientific theory is a shorthand encapsulation, in linguistic form, of the sensory experiences a practitioner using the theory either has had or would have (depending on whether it is explanation or prediction at issue) in the attempt to interact with the entities and properties—the systems, if you will—that the theory is about. The positivists were radical empiricists in the spirit of the British philosopher David Hume, and they saw the anatomy of scientific theories through the filter of Humean principles. What a theory does is to capture in linguistic form the causal regularities that hold within a domain of phenomena. For this reason, the positivist model of scientific theories made much of the concept of **theoretical laws**, such as the Newtonian laws of motion, the laws of chemistry, and the laws of genetics. They took **laws of nature**, in this sense, to be representable by certain kinds of linguistic statements. Hence, it seemed appropriate to the positivists simply to treat a scientific theory as a linguistic entity, a set of sentences possessing certain special logical features. This is an approach to understanding science that seeks to capture it at its most mature and successful: when it has evolved to the point of having a mature theoretical basis. Theories, with this approach, are after-the-fact reconstructions which present an abstract picture of the phenomena relevant to the domain. The positivists were interested in the final end-product of scientific inquiry more than they were interested in the sometimes sloppy process of trial and error used to reach that final end-product. We shall see that the positivist model of that end-product was conceptually rich but fraught with technical problems from the very beginning.

2.1 Theories as Formal Languages

Logical Positivism was a twentieth-century philosophical movement primarily concerned with theory of knowledge (epistemology), **mathematical logic**, and philosophy of science. Positivism was born in central Europe, but the flight of European intellectuals away from the growing menace of Nazi fascism in the years immediately preceding World War II spread positivism's influence across the larger part of the English-speaking world. Many of positivism's founders were initially trained as physical scientists before they became philosophers. For example, Rudolf Carnap was trained as a physicist, but he eventually came to occupy a position of enormous influence and respect within the positivist movement.

The logical positivists had a peculiar "spirit," an overriding set of intellectual and philosophical attitudes, preferences, and commitments. They got the second part of their name from the work of the nineteenth-century French philosopher August Comte. Comte had speculated that human intellectual culture evolves through three stages: theological, metaphysical, and positive. In the theological stage human science is at its most primitive, awash in religious superstition and busy attributing magical personlike properties to what are actually inanimate processes in nature. The world is populated by deities of various kinds who are held to cause what happens. In the metaphysical stage the magical gods are replaced by impersonal forces such as gravity and charge. This is still an error, according to Comte, but a better sort of error because it deanthropomorphizes the universe. Nevertheless, these forces are conceived as mysterious and occult in nature, "hidden" away "inside" things. The third and final stage is the positive stage in which the metaphysical forces are done away with in favor of commonsense observational concepts. The positivists followed Comte in being generally antireligious. They also advocated the demystification of nature. They adopted a radically empiricist orientation toward most philosophical issues (don't try to "think" the answer via metaphysical insight, let sensory experience decide the question, whatever it is). Many of them doubted the existence of disputed abstract entities (such entities don't really exist), or else they rejected outright all metaphysics, taking over from Comte the practice of treating 'metaphysics' as a dirty word in philosophy. They practiced a healthy skepticism toward the inflated claims made by theorists in many special fields of inquiry. Most important of all for our purposes, the logical positivists had a specific methodology that they employed in the analysis of philosophical problems, a methodology that explains the first half of their name, one in which mathematical logic played the lead role.

Mathematical logic was developed originally by Gottlob Frege and Bertrand Russell over a period spanning roughly from 1870 to 1915. Why were the positivists so smitten with mathematical logic? The answer is, because they understood mathematical logic to be a universally applicable, maximally precise cognitive tool, a tool that could be used to analyze successfully a host of abstract, difficult-to-understand concepts, problems, and issues. And a successful analysis ends our mystification about whatever it was that puzzled us by showing us how to understand it in terms of other things that we already understood previously.

The universal applicability of mathematical logic is due to the fact that it is a *subject-neutral language* that can be used to capture and express the claims of any special subject matter. And the precision of mathematical logic is due to the fact that the main symbols of this subject-neutral language can be provided with *mathematically precise and unambiguous meanings*. In short, you can use mathematical logic to talk about anything under the sun, and the claims you produce with it each have a unique and computable meaning. This is a splendid outcome because, as generality and precision increase, confusion and ambiguity decrease. It is no surprise then that this relatively new "toy" of mathematical logic was taken by the positivists to be a kind of philosophical wonder drug with almost Herculean powers.

The commitment to the power and predominance of mathematical logic was especially influential in positivist philosophy of science. The core assumption in positivist philosophy of science was this: Mathematical logic contains enough power and precision to allow us to capture and express in a useful and understandable way the most important causal and explanatory relationships that occur out in the real world, that is, in the domain of investigation in which a given scientist works. An example will help here. For almost 400 years since the Early Renaissance, philosophers of science (and philosophers in general) have been fond of asserting that the events that take place in the natural world are organized and controlled by laws of nature. The laws of nature that are operative within a given area of the natural world are perhaps the single most important feature of that area to the scientific discipline that investigates it. This was certainly true for the positivist model of science; for, under the positivist model of science, laws of nature had two important roles to play. First, laws of nature are what fix the causal relations in a system under scientific study: They are responsible for the relations of causation which hold between the subparts of a system and between that system and any other systems. Second, laws of nature make possible the explanation of natural phenomena. A change of properties in a system is explained by showing how that change was necessitated by the relevant laws of nature acting on the initial conditions in the system. We will investigate the topic of explanation in science in chapter 6. What is of interest to us here is the fact that laws of nature play such an important role in science, and yet saying just what a law of nature is proves to be rather difficult. Is a law of nature a command of God's? Or is it a mysterious connection of "necessity" between material properties? Prior to the rise of the positivist model of science no philosopher of science had succeeded in offering a clear analysis of the concept of a law of nature. The only clear conception that the metaphysicians could supply was theological, but a theological account of laws of nature was totally inconsistent with the spirit of positivism noted above.

The positivists had a solution to the puzzle, a solution that assumed that mathematical logic could rescue us from the vagueness surrounding the notion of a law of nature. The positivist solution was to treat laws of nature as representable by **material conditionals** of universal scope. A material conditional is a symbolic sentence in mathematical logic that has a specific logical form. In logic textbooks it is standard practice to claim that the material conditional represents what in English we say by using an 'If . . ., then . . .' pattern of speech. 'If A, then B' is a material conditional in which the truth of the component statement A is a condition sufficient for the truth of the component statement B. Let us consider an example from our discussion of the immunological theory of allergic disease in chapter 1. Recall that a vital stage in the generation of an allergic reaction is the physical binding of the allergen molecule to the antigen-binding portion of E-class antibody molecules, otherwise known as immunoglobulin E (IgE). Immunoglobulin E is one of five general classes of immunoglobulin antibodies found in the mammalian immune system. Immunoglobulin-E molecules have the same general three-dimensional structure as molecules of immunoglobulin classes A, D, and G, except for one notable difference: IgE molecules each have five distinct amino-acid regions in their heavy chains, whereas molecules of the other three classes of immunoglobulins mentioned each have four distinct amino-acid regions in their heavy chains. To oversimplify, IgE antibodies are structurally longer than either IgA, IgD, or IgG molecules. This is a law of immunology. It is structurally essential to IgE molecules that they each have five-component rather than four-component heavy chains. The positivist model of science would analyze this law of immunology as a universally quantified material conditional; that is, it would translate the English version of the law,

All IgE molecules have five-component heavy chains

into the symbolic sentence of first-order mathematical logic,

$$(\forall x)(E^{ig}x \supset H^{5}x)$$

where '$\forall$' is the quantifier symbol indicating the universality of the law (the universal quantifier), ' $\supset$ ' is the standard logical symbol for the material conditional ('if . . ., then . . .'), 'E^{ig}' is the symbolic term that refers to the property of being an IgE molecule, and 'H^5' is the symbolic term that refers to the property of having five-component heavy chains. The idea here is that the symbolic statement is a precise logical equivalent of the English one above it. The symbolic statement would be read in English as 'For any object in the entire universe call that object x, if x is an IgE molecule, then x has five-component heavy chains'.

To take another example from chemistry, the law that all pure metals conduct electricity can be rendered symbolically by the universally quantified conditional

$$(\forall x)(Mx \supset Cx)$$

in which 'Mx' is the predicate 'x is a pure metal', and 'Cx' is the predicate 'x conducts electricity'.

Now one might justifiably ask why this is something to get excited about. The positivist can answer that because we understand with mathematical precision the meaning of the material conditional ' $\supset$ ' and the meaning of the universal quantifier ' $\forall$ ', we now know with much more precision than we would have if we stayed with ordinary natural English just what the relationship is between being an IgE molecule and being an object with five-component heavy chains: It is the relation represented logically by ' $\supset$ '. We have arrived at the central assumption behind the positivist model of science. *The positivists assumed that the meaning of logically defined symbols could mirror or capture the lawful relationship between an effect and its cause.* Or, to put it in more general terms, the positivists assumed that logical relations between formal symbols could accurately represent external relations between the objects, properties, or events to which the formal symbols referred.

Now that we have gotten our feet wet in the sea of logical positivism, we are ready to dive in and take a look at the full positivist model of science. As we discussed above, the positivist model took theories as its unit of analysis in philosophy of science. Science is a collection of theories, according to the positivist, so we will want to see the positivist model of a scientific theory. A scientific theory is a formal language, a language expressed in first-order mathematical logic, which meets certain special conditions. First, the purely logical relations that are expressible in the language are the five standard truth-functions,

symbol	rough English meaning
v	or
&	and
$\supset$	if ...,then ...
$\sim$	not
$\equiv$	... if and only if...

plus the two standard quantifiers,

$\forall$	for all objects ...
$\exists$	there is at least one object ...

and the identity sign needed to assert that two referring expressions refer to the same thing,

$=$ ... is identical to ...

Second, the theory may contain any number of purely mathematical symbols such as 'Σ' that may refer to various constants (Planck's constant, Avogadro's number, and so forth). Third, all remaining symbols in the theory are neither purely logical nor purely mathematical symbols. They are what we might call substantive symbols, symbols like 'E^{ig}' which refers to the property of being an IgE antibody molecule. Fourth, these substantive symbols are divided into two completely separate kinds. Some of the substantive symbols refer to observable objects, properties, or events, and they are called observational terms. The remaining substantive symbols refer to nonobservable—more commonly called theoretical—objects, properties, or events, and they are known as theoretical terms. Objections have been made to this manner of speaking, because the proper contrast term to 'observational' is presumably 'nonobservational' rather than 'theoretical'; but we will stick with the standardly used terms throughout this text, thereby avoiding both a dispute over a technicality and terminological confusion. Fifth, the core content of a scientific theory is contained in the set of its purely theoretical laws of nature. These will be universally quantified conditionals whose only substantive symbols are theoretical terms. Sixth, but most important for our present purposes, the positivist model required that *every theoretical term in a scientific theory must be provided with an explicit definition composed entirely of observational terms.* Such explicit definitions were called **correspondence-rules**, or c-rules. For example, a positivist allergist has no business talking about the $F(ab')_2$ portion of an antibody molecule, which is the Y-end of it capable of binding antigen (the general term for any molecule bindable by an antibody), unless it is possible to provide an explicit definition of the theoretical term '$F(ab')_2$' in which the defining symbols are all observational terms. Now, to require that the definition be an *explicit* one is to require that the observational side of the defintion be such that whenever it truly applies to an object, property, or event, its application is necessary and sufficient for the theoretical side of the definition to be truly applied to that object, property, or event. Put another way, for two terms to be **necessary and sufficient conditions** for each other means that whenever the one term applies to a phenomenon so does the other, and whenever one of the terms does not apply to a phenomenon neither does the other. The logical relation that holds between two terms when each is necessary and sufficient for the other is represented by the logical connective called the **material biconditional**, ' ≡ '. So, our positivist allergist must provide a c-rule for the theoretical term '$F(ab')_2$' that has a particular logical form: It is a symbolic sentence in first-order logic that is a universally quantified material biconditional.

Suppose our positivist allergist decides to define observationally '$F(ab')_2$' in terms of a laboratory procedure combining a pepsin digest with gel immunoelectrophoresis. This standard laboratory procedure can be performed on blood serum to observe the various fragments or portions of antibody molecules, including the $F(ab')_2$ fragment. If the immunologist makes no mistakes in the procedure, the $F(ab')_2$ portions of the antibodies show up as visible arclike bands against a uniform background at the end of the procedure. In Figure 2.1 I have depicted what the result of this procedure would look like. The figure shows an immunoelectrophoresis gel in which an antigen well and an antibody trough have been cut. The arc to the right of the well between the well and the trough was caused by the reaction of the $F(ab')_2$ fragments with their corresponding antigen. Those deeply theoretical objects have made contact with human observational space at that endpoint in the procedure. So, says the positivist allergist, let us simply define '$F(ab')_2$' in terms of this procedure that results in the observation of the visible arcs. Let us represent the observational procedure used to generate the bands by the formal symbol 'O^1'. Then the c-rule for '$F(ab')_2$ would look like this:

$$(\forall x)(F^{(ab')2}x \equiv O^1x)$$

In ordinary English the above symbolic sentence translates as, 'anything is an $F(ab')_2$ portion of an antibody if and only if it produces observable arcs of a certain kind at the end of observational procedure O^1'.

The ability to provide c-rules in the form of explicit definitions served a doctrinaire purpose for the positivists. Many positivists claimed that the c-rule requirement was what kept science "honest" in the sense that it served to inhibit scientists from arbitrarily making up exotic theoretical objects, properties, or

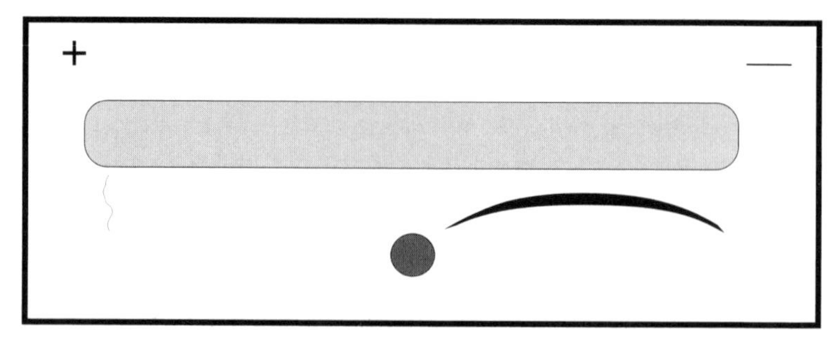

Figure 2.1. The observable outcome of the immunoelectrophoresis of a G-class antibody pepsin digest.

events. The c-rule requirement therefore was one main feature of a scientific theory that distinguished genuine scientific theories from their phony imitators. A phony science is called a **pseudoscience**, and many such pseudosciences are big businesses in contemporary life. Examples are not hard to find. Astrology and psychic healing, for example, are big moneymakers for their practitioners. One interesting aspect of pseudoscientific theories is that they contain no c-rules for their pet theoretical terms. It is unheard of for an astrologer to provide a c-rule for 'astral influence' or for 'influence of the ascending house'. The c-rule previously provided for '$F(ab')_2$' is admittedly very esoteric and unfamiliar to those unacquainted with immunological science, but at least it can be provided. That is more than can be said for the typical pseudoscientific theoretical term.

The reader might now ask why it is such a negative feature of a theory that it cannot provide c-rules for its theoretical terms. Why couldn't the pseudoscientists defend their theories with a simple shrug of their shoulders and a cavalier "so what"? The reason why such a defense would be lame, at least according to the positivist, is because *c-rules tell us how the alleged theoretical entities defined by them make a difference to observational experience.* If you cannot find a difference that the alleged existence of the theoretical object, property, or event makes to the observable world, then the positivist would suggest the absence of any good grounds for supposing that such theoretical objects, properties, or events exist at all. The positivist model tied the grounds for asserting the existence of a thing to the difference its existence would make to our observational experience. Suppose a controversial immunologist proclaims the discovery of a new class of theoretical objects that are active in the mammalian immune system at an extremely fine level of detail. This immunologist invents a new theoretical term to name these objects: 'ferple'. Are there really ferples? we wonder. So we question this immunologist about ferples only to find that ferples are *so* darn theoretical, *so* darn small, and so on, that they can never be detected by any observational means. They exist, insists the maverick immunologist, but we can never observationally experience their presence or any of their direct causal effects. The only causal effects ferples have that we can observe are so far downstream from ferples themselves—involve so many intermediate causal links—that it is just as likely that ferples are not really the first link of that chain of causes and effects as it is likely that they are the first link. The problem is not that there is currently no possible evidence *for* the existence of ferples; as we have imagined the case, the problem is that there is no possible evidence that would count *against* the existence of ferples. Now if we are positivist philosophers of science, we must reject the existence of ferples. They make no difference to the way the world can be observed to be by us, so we have no warrant for supposing they exist. To put it bluntly, if they make no difference to observational experience, then they aren't there! Another way to understand the point the positivist is making is to consider what our estimate would be of the scientific worth and meaning of a theoretical term whose use is consistent with any and all possible observational experiences. The positivist would argue that such a term is scientifically meaningless. It is important for the reader not to misunderstand the positivist here. The positivist is not arguing that science shouldn't be allowed to postulate theoretical entities; the positivist is saying only that any such theoretical entities must be given an explicit market value within observational experience.

Requiring that theoretical entities have an observational market value was supposed to help reduce the number of bogus theoretical entities in science. Whether the requirement was ever capable of achieving that outcome is doubtful, but we can at least understand how it was supposed to work in broad outline. The history of science contains examples of theoretical entities and properties that turned out in the course of inquiry not to exist at all. For example, early in the history of chemistry the combustion of a chemical substance was postulated to be due to the release of something called phlogiston, a mysterious "principle" of pure flammability often conceived to be a substance of some sort that all combustible bodies contained. There is no such substance or principle. In physics, once upon a time, space was held to be everywhere pervaded by a mysterious nonmaterial substance called the aether. Aether was the alleged "medium" through which energy propagated and ordinary matter moved. The case of aether is justly famous in the philosophy and history of science for being at the center of the overthrow of Newtonian physics at the beginning of the twentieth century. Again, there is no such aether—at least there is not if Einsteinian physics is correct. The positivists were aware of such historical cases of bogus theoretical entities and hoped that the c-rule requirement in their model of science would contribute somehow to lessening the frequency of bogus theoretical terms; for, if necessary and sufficient conditions expressible solely in the observational vocabulary are required for every theoretical term, the stringency of such a requirement would surely serve as a deterrent to the casual postulation of theoretical entities.

2.2 A Dummy Scientific Theory to Illustrate the Positivist Model

The discussion of the positivist model of scientific theories up to now has been a bit abstract. A specific example would help us to understand how the model was supposed to work. Let us construct a crude and completely phony positivist theory of allergic disease so that we might illustrate how the positivist model was meant to work. I will call this theory a "dummy" theory because it is completely made up and because it bears the same sort of relation to genuine immunological theories of allergic disease that human-shaped dummies (like the dummies used in car-collision tests) bear to living humans. What we lose through the theory's phoniness we gain back from its simplicity and clarity. We will be able to see all the components of the positivist model in sharp relief.

Our dummy theory of allergic disease contains four substantive theoretical terms, three theoretical laws of nature, and four c-rules. We will allow the theory to contain any number of additional observational terms, and we shall assume for present purposes that telling the difference between theoretical terms and observational terms is unproblematic (although, as we shall see in chapter 3, this is a very questionable assumption). Since the positivist model took the core component of a scientific theory to be its laws of nature—especially its purely theoretical laws of nature—we will begin with the three laws in the dummy theory of allergic disease. I will present each law in both its English and its symbolic form. The laws will be numbered L_1 through L_3.

English		Symbolic
L_1:	All mammals with gene G have serum blickledorps	$(\forall x)(Gx \supset Bx)$
L_2:	All mammals with serum blickledorps secrete microspam	$(\forall x)(Bx \supset S^m x)$
L_3:	All mammals that secrete microspam experience allergic symptoms	$(\forall x)(S^m x \supset Ax)$

The properties of being a mammal with gene G, having serum blickledorps, being a mammal that secretes microspam, and being a mammal who experiences the symptoms of allergy, represented symbolically by the theoretical terms 'G','B', 'S^m', and 'A', respectively, are all theoretical properties. It is important to notice that I have deliberately chosen to use a couple of phony theoretical terms like 'blickledorps' and 'microspam' (there actually are no such things) to emphasize by exaggeration their theoreticalness. This is in the spirit of positivism. As far as the positivist is concerned, you don't need to know what 'blickledorp' and 'microspam' mean on a conceptual level (that is, their sizes, colors, charges, polarities, and such, if any); you only need to know the meanings of the *logical* relations that hold between 'blickledorp', 'microspam', and any other terms, be they theoretical or observational, with which they are connected by the laws of the theory. Here we see hard at work the positivist assumption mentioned previously, the assumption that logical relations are capable of accurately representing causal connections out in the real world. For example, according to L_2 a mammal with serum blickledorps is an entity that necessarily secretes microspam. According to L_1, having serum blickledorps is a causally necessary result of the possession of a certain gene. How do we know this? The positivist answers: because the logical relations between the linguistic symbols 'G', 'B', and 'S^m' in laws L_1 and L_2 represented by the material conditional ' $\supset$ ' accurately capture or mirror the objective causal connections between physically four-dimensional genes, blickledorps, and microspam molecules. To the positivist, the logical relation denoted by ' $\supset$ ' is less cognitively remote and better understood (because it is precisely and unambiguously defined in mathematical logic) than the objective causal connections themselves; so, we can be more sure of the logical relations than we can of the objective causal connections. It is a question the reader should ask at this point, whether this general assumption by the positivist that *logical connectives mirror causation* is warranted or not.

It remains for us to complete our dummy theory of allergic disease by providing c-rules for our four theoretical terms. Once again, I will simply make them up so as to emphasize through exaggeration the difference between the theoretical half of a c-rule and the observational half of a c-rule. To facilitate understanding, I will leave the theoretical side of each c-rule in its formal version but present the observational side of each c-rule in English.

1. $(\forall x)$ ($Gx \equiv$ x's fingernails produce green spots when pressed continuously for five minutes)
2. $(\forall x)$ ($Bx \equiv$ pricking x's skin immediately produces a bright turquoise spot on the surface of x's tongue that lasts four seconds before fading away)
3. $(\forall x)$($S^m x \equiv$ x's breath smells like maple syrup)
4. $(\forall x)$ ($Ax \equiv$ x presents with the typical symptoms of allergies—wheezing respiration, watery eyes, swollen and inflamed nasal tissues with pink coloration, hives, and so forth)

We should be clear that the right side of each of these c-rules (the side in English) is to be taken as a purely observational description.

Now that we have our illustrative dummy theory completed we may ask to be shown what benefits flow from logically formalizing a theory in such a manner. The positivist answers that there are important benefits forthcoming. Perhaps the most prominent benefit is that *theoretical inferences become a matter of making deductively valid inferences*. This makes such inferences completely clear, precise, and transparent to the practitioners who use the theory. Suppose, for example, that a certain patient attended by our positivist allergist is examined by the allergist and tested for blood blickledorps. This patient's blood is found to contain no blickledorps (assume the test for serum blickledorps is extremely accurate). Now, by law L_1 and the deductive inference rule logicians call *modus tollens* (which stipulates that from a conditional statement 'if A, then B' and the statement 'not B', we can validly infer the statement 'not A'), the allergist can infer deductively that the patient does not have gene G. So what?, you say, why is that a big deal? The positivist answers that it is a big deal because this inference is not a guess, not a vague hunch or prediction; rather, it is a statement the allergist can proclaim with *deductive certainty*. To say that a pattern of inference is deductively certain is to say that *if* the premises (the supporting statements, in our example they are the two premises L_1 and the statement 'the patient's serum contains no blickledorps') are true, *then* the conclusion (the inferred statement, in our example it is the statement 'the patient does not have gene G') *must* be true also. This alleged elimination of guessing from theoretical inference is one major advantage the positivist claims for the positivist model of scientific theories. Further, the *modus tollens* inference counts as a double-check against any independent test for having gene G; that is, we can infer the lack of gene G without having to administer any independent tests for G to the patient.

A second major advantage of logically formalizing a theory should perhaps be mentioned at this point, although we shall investigate it much more thoroughly in chapter 6. Under the positivist model, an especially powerful and precise model of scientific explanation becomes available to us. Using our dummy theory, for example, the appearance of the symptoms of allergy in a human patient could be scientifically explained in the sense that a statement describing the patient's symptoms can be logically deduced from L_1, L_2, and L_3 together with an additional pertinent premise that attributes to the patient in question the gene G mentioned in L_1. Explanation in science, claimed the positivists, consists in showing how the event to be explained was necessitated by the relevant laws of nature together with the relevant initial conditions obtaining in the system in question. Thus, explanation in science is put on a solidly precise footing. Scientific explanation need not remain the largely impressionistic and vague notion it was and often remains outside the positivist model.

The positivist model of scientific theories dominated philosophy of science for nearly forty years after its introduction in the 1930s. The precision and power of the positivist model silenced both its critics outside the positivist movement as well as any self-doubters within the positivist camp. There was a general consensus among philosophers of science about the nature of science, a broad set of common beliefs and assumptions about the methodological, epistemological, and ontological features of mature scientific theories. Nevertheless, minor "technical" problems with the positivist model soon began to plague its devotees. In the course of trying to iron out the wrinkles in the model the positivists soon discovered that the technical problems weren't so minor after all. What began as nitpicking about small errors of detail soon grew into the general worry that the positivist model really could not do much of the work it was designed to do. There was much trouble in positivist paradise, and we shall examine closely the nature of that trouble in the next chapter.

Further Readings

A famously clear presentation of the positivist model of science is found in F. Suppe's Introduction to *The Structure of Scientific Theories*, 2nd. ed., edited by F. Suppe, Urbana, Ill., University of Illinois Press, 1977. A late version of the positivist model can be seen in action in a variety of different domains in H. Kyburg, *Philosophy of Science: A Formal Approach*, New York, Macmillan, 1968.

3

Trouble in Positivist Paradise

The distinction between observational terms and theoretical terms is a critically important component of the positivist model of scientific theories. Without a workable distinction between the observable and the theoretical the central demand which the model makes that every theoretical term be provided with an explicit definition containing only observational terms would be a bit perverse, for it would be an exercise with no point to it. Initially, few positivists took much care to examine closely this distinction that played such a vital a role in the positivist model. Rudolf Carnap, one of the original and towering figures of logical positivism, helped himself to the observational/theoretical distinction without providing a terribly precise or useful criterion for telling the difference between terms that refer to observable phenomena and terms that refer to unobservable (theoretical) phenomena. We will begin this chapter by looking closely at a number of different criteria designed to support the observational/theoretical distinction.

3.1 The Point of the Observational/Theoretical Distinction

A *criterion* for distinguishing observational terms from theoretical terms is a set of conditions, which, if they are satisfied, qualify a term as referring to observable objects, properties, or events, and not to theoretical ones. But there are good grounds for holding that an imprecise and vague criterion for distinguishing observational terms from theoretical terms would be a seriously detrimental problem for the positivist; for the observational/theoretical distinction is intended to carry methodological, epistemological, and ontological significance. That is, something observable is something that we can come through *systematic investigation, to know, exists.* Therefore, there is no special problem about the legitimacy of observational terms in a scientific theory. On the other hand, theoretical terms, by the nature of the case, refer to regions of reality less accessible to direct human experience. The history of science is filled with theoretical terms that turned out not to refer to anything at all. Terms like 'phlogiston' 'aether', to take the examples from the history of chemistry and physics we previously noted, do not refer to anything that exists. Within immunology itself, one could argue that Paul Ehrlich's theoretical term '**horror autotoxicus**'—a term Ehrlich used to refer to the alleged inability of the immune system to mount an immunological attack against an organism's own tissues (which we know to be no inability at all because of the discovery of autoimmune diseases such as multiple sclerosis)—is just such a bogus theoretical term. Such examples emphasize the naturally greater worry about the legitimacy of theoretical terms that every experienced scientist will acknowledge. Because the observational/theoretical distinction is required to perform the important methodological, epistemological, and ontological labor that we have just revealed, it is all the more curious that the distinction was taken for granted by most positivists.

3.1.1 Ease of Application

Carnap's criterion for distinguishing observational terms from theoretical terms strikes a contemporary philosopher of science as being far too casual. It is so nonchalant as to be jarring.

> Carnap's criterion: A term *t* is observational if a practitioner using the theory in which *t* occurs could determine *t*'s correct application with relative ease, that is, quickly and with minimal effort; otherwise, *t* is a theoretical term.

Our immediate question is, ease of application relative to what? The positivist might answer, ease of application relative to the greater difficulty and complexity of determining the correct application of a theoretical term. Carnap admitted that this criterion made the observationality of a term dependent on and relative to the level of knowledge and experience of the practitioner who uses it. This can be illustrated with an example. Some allergists claim to be able to immediately diagnose allergic disease in a new patient by noting the color of the patient's nasal and throat membranes. Allergically inflamed tissue, as opposed to infectiously inflamed tissue, is flooded at a microscopic level with white blood cells called eosinophils, which give the tissue a characteristically pinkish tint, and which a trained and experienced eye is able to contrast with the deeper reddish color of infected tissue. Now, does this mean that 'eosinophilia'—the technical term allergists use to refer to the situation in which there is an overabundance of eosinophils in a local region of tissue—is an observational term in immunological medicine? The answer would seem to be no. What the allergist observes is the color of the inflamed tissue, not the eosinophilia associated with the color, for eosinophilia is a microphenomenon causally connected to the color the allergist observes. Yet, it is uncontroversial that an experienced allergist can correctly apply the term 'eosinophilia' quickly and with minimal effort (a quick look at the patient's upper respiratory membranes is all that is needed). But of course only complicated laboratory tests that presuppose all kinds of theoretical laws of nature could confirm the presence of many eosinophils in the patient's upper respiratory tract tissues. In fact, to be technically precise we should note that 'eosinophilia' refers to an overabundance of eosinophils *in the blood*, at least, that is the term's original meaning. The extension of the meaning of the term to cases involving tissue membranes by allergists adds yet another layer of theoretical inference to the situation. In short, Carnap's criterion would make too many theoretical terms (terms like 'eosinophilia') observational. It *misclassifies* clear cases. No criterion that misclassifies clear cases is acceptable.

In general, ease of application is a poor indicator of whether the phenomenon that a term allegedly refers to is observable or unobservable; for downstream causal effects of an unobservable (theoretical) entity can be used by an experienced practitioner to apply correctly, quickly, and with minimal effort, the term referring to that theoretical entity.

3.1.2 Instrumentation

The problem that Carnap's criterion suffered from might be thought to point the way to a better criterion. Because we need artificial instruments of various kinds—microscopes, chemical stains, slides, and so on—to confirm the presence of eosinophilia in a patient, perhaps the general lesson to be learned is that the use or nonuse of artificial instruments has something to do with distinguishing observable phenomena from nonobservable phenomena. Because the positivist model treated science as composed of theories, and because it treated theories as linguistic things (organized pieces of mathematical logic), once again we must express our new criterion as one that distinguishes into two classes the terms of a scientific theory.

> The instrument criterion: A term *t* is observational if a practitioner does not require the use of an artificial instrument to apply or use *t* correctly; otherwise *t* is a theoretical term.

One can easily see how this criterion is supposed to work. A theoretical term is one whose correct application *does* require the use of an artificial instrument. If you need to use an artificial instrument to apply the term correctly, then the term must refer to phenomena you cannot observe. This has an intuitive appeal to many people, which we might put by saying that,

> if you need a newfangled doohickey of a device to detect, measure, or manipulate something, then that something is theoretical, if anything is.

The mammalian immune system contains an interesting type of cell known as a natural killer, or NK, cell. NK cells are capable of killing other cells that have become cancerous or have become transformed by viral infection. NK cells are also capable of "identifying" and killing cancer cells and virally infected cells without ever having been exposed to such cells before (that is the point of calling them *natural* killer cells). Clearly, 'NK cell' is a theoretical term in immunology; for to detect, measure, or manipulate such cells artificial instruments must be used. For example, to detect the presence of NK cells in the blood or tissues of a mammalian organism immunologists typically perform a ^{51}chromium release assay on a specially prepared sample of the blood or tissue, a procedure that requires the use of a scintillation spectrometer to measure radioactive radiation. By contrast, 'wheal' is an observational term in immunological medicine, for it refers to a raised, red local inflammation of the skin (a welt or a hive, in nonmedical terms), and detecting or measuring such inflammatory eruptions of the skin does not require the use of artificial instruments by the practicing immunologist. Or does it? Suppose our positivist allergist is testing a new patient using standard skin scratch tests. A small amount of a suspected allergen is placed into a tiny hole in the patient's skin that was made with a sterile pin point. Our allergist waits about twenty minutes and then examines the skin of the patient. If the patient is allergic to the particular suspected allergen a wheal eruption should be visible where the pinprick and allergen were administered; that is, the skin at that location should be raised and red. Allergists measure the rough degree of allergy by *how* raised and red the wheal is. But suppose our positivist allergist wears eyeglasses or contact lenses and is nearly blind without them. In that case, the allergist will need the glasses or contact lenses to detect and measure accurately the presence of any wheals on the patient's skin. If so, then the allergist will need to use an artificial instrument to apply or use accurately the term 'wheal'. It would follow under the instrument criterion that 'wheal' is a theoretical term, not the observational one we know it to be. The positivist cannot escape this unfortunate result by appealing to statistically normal practitioners, for the statistically normal allergist may require artificial eyewear to see properly. So it looks as though the instrument criterion misclassifies clear cases just as Carnap's criterion did.

We can identify a general problem for the instrument criterion from this unfortunate misclassification of 'wheal' as a theoretical term. The general problem is that the instrument criterion suffers from a vagueness inherent in the notion of an artificial instrument. If we adopt a broad notion of what it is to be an artificial instrument, a broad reading under which eyeglasses and contact lenses count as artificial instruments, then something as obviously observable as the wallpaper pattern on a wall a few feet in front of a nearsighted person who is without his or her glasses or contact lenses counts as a theoretical aspect of the universe. That is surely absurd. Such a broad reading of 'artificial instrument' would make all kinds of observable objects, properties, and events theoretical.

On the other hand, how narrow should the notion of an artificial instrument be trimmed to avoid the misclassifications? Some simple ways of narrowing the notion are clearly not going to work. For example, it would not do to alter the instrument criterion to read, a term *t* is observational if no *expensive* artificial instrument is necessary to apply or use *t* accurately. There is no justification for holding the cost of constructing an instrument to be a reliable indicator of the theoreticalness of what the instrument detects, measures, or manipulates. My late uncle once paid almost $2,000 for a pair of special eyeglasses that make it possible for people with an eye disease called macular degeneration to see everyday objects in their environment which persons without macular degeneration can see unaided. Neither will it do to alter the instrument criterion so that the *complexity* of the artificial instrument is what counts, for there is no general way to answer the question of how complex an artificial instrument must be before what it helps a scientist detect, measure, or manipulate is rightfully considered to be a theoretical phenomenon. What one person finds imposingly complex in machinery another person finds simplistic and primitive, even where all the parties are trained practitioners. But surely, if observationality is to carry the kind of methodological, epistemological, and ontological burden which we saw previously it is intended to carry, then it cannot be a status a term has merely because of what happens to amaze or impress some practitioner as being a real doohickey of a device.

The instrument criterion does get one thing right about the observational/theoretical distinction, as we can see by asking the question, why would a practitioner need a complicated instrument to detect, measure, or manipulate an object, property, or event? The commonsense answer is, because that object, property, or event is located at a level of organization in the size scale of the universe at a significant "distance" from the level of organization to which our sensory receptor surfaces can have access unaided by fancy instruments. Theoretical terms are those terms that refer to objects, properties, or events that are either very small or very large relative to the objects, properties, and events to which our unaided biological senses can have normal functional access. If this commonsense intuition is correct, then the instrument criterion is a natural aftereffect of the real source of the observational/ theoretical distinction: the fact that we humans occupy a limited size niche in the size scale of the universe. But the instrument criterion incorrectly attributes the observationality of a term *t* to the fact that an instrument is required to gain access to the phenomenon referred to by *t*, when the genuine ground of *t*'s observationality is due to the size-scale remoteness of the phenomenon referred to by *t*, a remoteness that makes the phenomenon's direct observation problematic for creatures with our biological structure. We will return to this point in section 3.1.5.

3.1.3 Putnamian Cynicism

The recognition of the fact that the instrument criterion is inherently vague and imprecise led some philosophers of science to the cynical conclusion that the observational/theoretical distinction was invented to solve a problem that in fact does not exist. Hilary Putnam argued that no observational/theoretical distinction can be drawn in any nonarbitrary manner; for if an observational term is one that in principle can *only* be used to refer to observable objects, properties, and events, then, because any term *could* be used under some circumstances to refer to *un*observable objects, properties, and events, *there are no observational terms at all.* For example, the color term 'white' might be thought to be a solidly observational term, if anything is; yet, 'white blood cell' is presumably a theoretical term given the relatively complicated instrumentation needed to photograph, count, or type white blood cells. The obvious problem with this argument of Putnam's is that there is no good reason to accept the conditional presupposition; there is no good reason to suppose that an observational term is one that *in principle* can *only* be used to refer to observational phenomena. In fact, a closer look at Putnam's argument reveals that it is a symmetrical argument; that is, the same argument would show that *there are no theoretical terms at all*, for there are no terms that *in principle* can *only* be used to refer to *un*observable objects, properties, and events. For example, 'particle' is surely a theoretical term as used by a subatomic particle physicist, if anything is; yet, 'dust particle' is an observational term given the everyday fact that dust particles are perfectly visible with the unaided eye in a shaft of sunlight.

The moral Putnam draws from all this is that the problem for which the distinction between observable entities and theoretical entities was invented is an illusion. There never was a time, Putnam argues, when people spoke only about observable states of affairs. In a famous passage, he writes,

> There never was a stage of language at which it was impossible to talk about unobservables. Even a three-year-old child can understand a story about "people too little to see" and not a single "theoretical term" occurs in this phrase. ("What Theories Are Not," p. 218)

If Putnam is correct, then there is no *special* epistemological worry about the legitimacy of theoretical terms, for they have always been with us, and their epistemological (as well as ontological) status is of a piece with that of observational terms. The suggestion is that the positivists wholly invented a special worry about the legitimacy of theoretical terms, after which they experienced a demand for a criterion to distinguish observational terms from theoretical ones.

The problem with Putnam's position is that, while a believer in theoretical entities and properties can cheerfully agree that observable and theoretical entities are on a par ontologically, the claim that they are on a par epistemologically would seem to beg the question against the doctrinaire positivist. The usual move here is to cite the "bad" historical cases—the cases of phlogiston and the aether, for example. The plain historical fact of the matter is that we have run and continue to run a higher risk of error about theoretical entities than we do about observable ones. No one who is familiar with the history of science could seriously dispute otherwise. Putnam's position far too casually dismisses that history and the legitimate worry about the epistemological status of theoretical terms that it generates. There *is* something to the greater risk of error about theoreticals, and so there is something to the lesser degree of confidence we have in postulating theoretical entities and properties. For this reason, before we opt for the cynical view of Putnam's, we should exhaust the available criteria for distinguishing between observational terms and theoretical terms.

3.1.4 Encapsulated Information Processing

Certain influential doctrines about language in general, ultimately traceable to the work of the renowned American philosopher W. V. O. Quine, led some philosophers of science in the 1950s and 1960s to conclude that the justification for making any claim about an observable phenomenon actually depends on the assumption that other conditions hold good, some of which involve highly theoretical matters. These other conditions are different from but related to the observable phenomenon in question. They are in fact so related to the observable phenomenon in question that, without assuming that they hold good, one cannot justifiably claim to have observed the observable phenomenon in question. For example, an allergist can assert that he or she has observed a negative response in a patient being tested for allergy to aspergillus mold *only if* the allergist assumes that a host of other "background" conditions hold good. Specifically, an intradermal skin prick test involving aspergillus mold spores that does not produce a wheal counts as a negative response only if the patient being tested is not temporarily on certain medications whose anti-inflammatory effects interfere with the dermal inflammatory response of mast cells in the patient's skin, and only if the aspergillus mold spores have not become so denatured during storage through being subjected to subfreezing temperatures that they no longer function like normal aspergillus mold spores, and only if the mast cells in the patient's skin are distributed normally so that there are enough in the particular location where the skin prick was made to produce a wheal should the patient be allergic to aspergillus mold, and only if . . . , and so on.

The notion that scientific claims and inferences are interdependent in this way is called **holism**, or the holistic model of scientific theories. We will examine this influential model of scientific theories in greater detail in chapter 4. For our present purposes, what matters is the implication such a holistic view of science has for the observational/theoretical distinction. The basic implication of interest is this: If holism is correct, then what something is observed to be like is a result not only of the intrinsic features of that something, but also a result of what other background beliefs the observer possesses, including some highly theoretical background beliefs. This implication is often expressed by the holistic philosopher of science in language charged with doctrinaire rhetoric: *There is no such thing as a theory-neutral observation-language, a language in which observational terms have meanings that are the same in every theory and that therefore can be used to report observable facts that hold good in every theory.* Such a view obviously is not compatible with an observational/theoretical distinction that packs a traditional epistemological punch.

The philosopher of science Jerry Fodor reacted against this holistic gutting of the observational/theoretical distinction by offering yet another criterion for making the distinction on a nonarbitrary and principled basis. Fodor argued that the difference between observable phenomena and theoretical phenomena rests on whether the claim involving the phenomenon in question was arrived at or "fixed" *relatively inferentially* or *relatively noninferentially*. Fodor argued that empirical evidence from experiments in perceptual psychology involving human subjects shows that the "information processing" that takes place inside of specific sensory systems—visual processing, auditory processing, and so on—is relatively isolated information processing. For example, the processing of neurological signals from the retinae of the eyes by the visual circuits in the brain is relatively isolated or cut off from influence by information being processed in other circuits of the brain. In particular, the signal processing in the five traditional sensory pathways (vision, hearing, taste, touch, and smell) is relatively cut off from information contained in higher, more central circuits of the brain. The sensory pathways are "out in the boondocks" where they remain relatively uninfluenced by what is happening in the "central cities" of the higher and more centralized brain pathways. Cognitive psychologists call such isolated processing "encapsulated" processing. Because it is reasonable to assume that our highly theoretical beliefs involve abstract reasoning circuits located in the more centralized information processing circuits, Fodor postulated that the information processed in the peripheral, encapsulated sensory circuits is information relatively unaffected by (and certainly not dependent on) our other theoretical beliefs; in short, it is more "observational" information. This suggests the following criterion for distinguishing observational terms from theoretical terms:

> The encapsulated information processing criterion: A term *t* is observational if it is used to describe phenomena detected through cognitive processing encapsulated in peripheral sensory receptor systems; otherwise *t* is a theoretical term.

Notice that Fodor is careful not to claim complete encapsulation, for that is the point of saying that the encapsulation is a relative one. *Some* very general theoretical beliefs presumably affect virtually every information processing system in the brain, so it is a matter of the degree of encapsulation.

The problem with Fodor's criterion is that it rests entirely on an empirical claim about the structure of the human brain that may turn out to be false: namely, that the information processing which takes place in peripheral sensory systems really is encapsulated to the extent that Fodor presumes. The evidence Fodor uses in support of his criterion is far from conclusive, and further empirical investigation could very well turn out to support the holists against Fodor. The philosopher of science Paul Churchland has argued that the empirical evidence from cognitive psychology shows that the encapsulation that Fodor requires does not occur in the human brain, that peripheral information processing is not *that* isolated from other more centralized information circuits in the brain.

It would not profit us to descend into the technical literature relevant to the dispute between Fodor and Churchland. What the positivist model of scientific theories needs is a basis for the observational/theoretical distinction that is not susceptible to being destroyed by what psychology eventually discovers about the cognitive structure of the human nervous system. If I am correct in my suggestion that the basis for the observational/theoretical distinction has something to do with the fact that human beings occupy a local size niche in the universe, then whether information processing in peripheral sensory systems is encapsulated should have little to do with whether 'E^{ig}' is a theoretical term or an observational term in immunological science. The greater epistemological "safety" of scientific claims involving only observational terms, and therefore the greater ontological "safety" of such claims, arises from our having *closer unaided sensory access to the observable*, not from the fact (should it turn out to be a fact) that neurally encoded information *about* the observable is processed within encapsulated sensory circuits.

3.1.5 Sensory Surface Distance

Consider Figures 3.1, 3.2, and 3.3 in order. To appreciate the point of these figures recall our investigation of the mechanism of allergic disease in chapter 1. The classical kind of allergic reaction, now called type I hypersensitivity, occurs

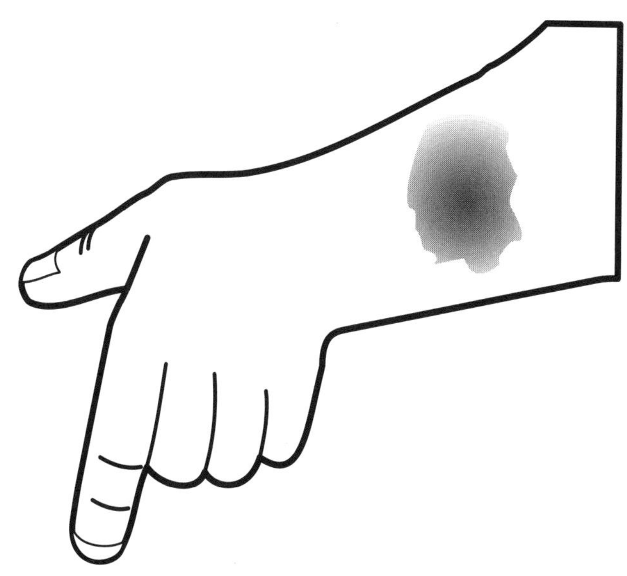

Figure 3.1. An inflammatory swelling from an allergic response.

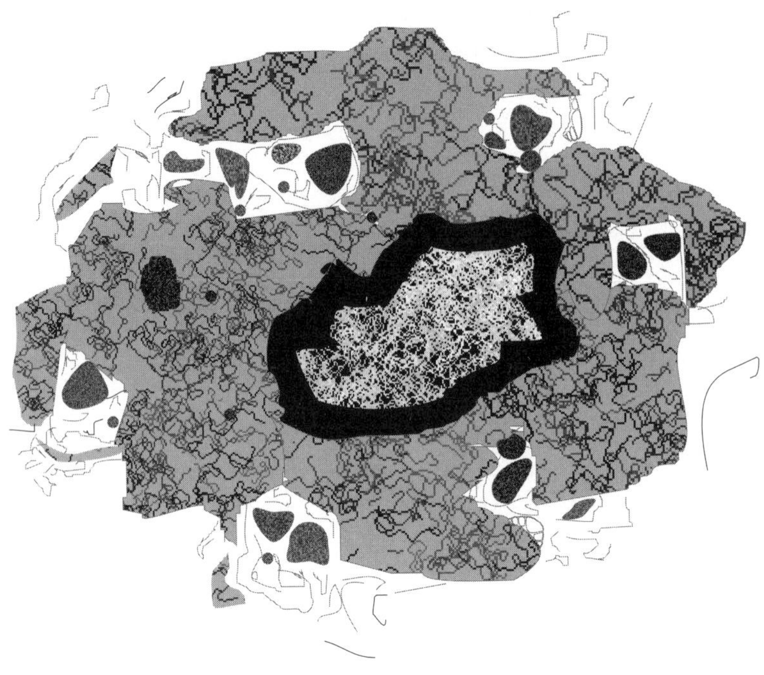

Figure 3.2. A degranulating mast cell.

when the offending substance (the allergen) binds to and cross-links tiny IgE molecules, which are themselves fixed to the outer membranes of mast cells. The binding of the allergen triggers a series of chemical events inside the mast cells which results in the rupture of their membranes and the release into the local tissue of the many granules that mast cells contain within their cytoplasm. These granules contain many preformed chemicals that produce inflammation such as histamine, as well as many chemicals that are components of inflammatory mediators subsequently synthesized in the inflamed tissue.

In numbered order the three figures depict events that move from the highly observational to the highly theoretical. In Figure 3.1 we can observe without the use of elaborate instruments the swollen and inflamed tissue of an allergic reaction to bee venom, a reaction called anaphylaxis. Figure 3.2 involves a region of reality that requires elaborate instruments to detect. We can take a photograph of a mast cell degranulating as seen through a powerful microscope, but only after we have chemically "prepared" the mast cell in a special way, perhaps through staining it with dyes, or through shrinking the cell's membrane onto its internal granules (and one hopes that one hasn't so distorted the real nature of the mast cell by these treatments so that what we see in the resulting photographs is not what the mast cell is really like). In another sense, we do not observe a degranulating mast cell at all; for what we literally see in such photographs are shapes and lines of gray and white like those depicted in Figure 3.2. We need to assume the truth of a host of background theoretical beliefs involving how

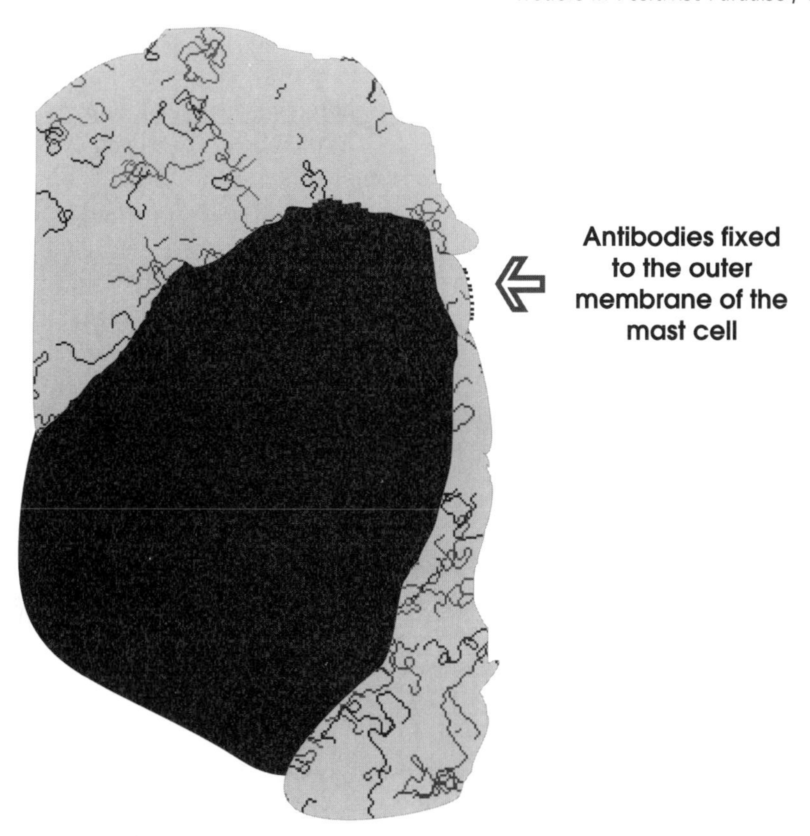

Figure 3.3. Membrane-bound E-class antibodies on the surface of a mast cell.

the machinery needed to make the photo works, and how the object viewed was prepared, to believe that the gray and white shapes and lines *are* a photo of a mast cell in the process of discharging its granules.

Lastly, consider Figure 3.3 where we arrive at microstructure in the immune system that is at an even greater "distance" from our unaided senses than is the microstructure involved in Figure 3.2. With an even more powerful microscope we can take a photo of an area of the outer membrane of a mast cell where there appear to be little gray wisps apparently growing out of the membrane. The wisps look like tiny hairs, just as I have portrayed them in the figure. If you accept a host of theoretical assumptions involving the machinery used to make the photo, as well as what had to be "done" to those wisps to make them visible at all (one technique is to tag them with colloidal gold particles), then you can accept the claim that those little gray wisps *are* E-class antibody molecules fixed to the outer surface of a mast cell.

We can summarize the lesson we learn from Figures 3.1, 3.2, and 3.3 in the following way. You don't have to make very many risky theoretical assumptions to observe swollen tissue to be swollen tissue; you do have to make a small number of modestly risky theoretical assumptions in order to interpret a bunch of gray and white shapes and lines in a photograph to be a degranulating mast cell; and you have to make quite a few moderately risky theoretical assumptions to interpret a bunch of tiny gray wisps in a photograph to be antibody molecules fixed to the outer membrane of a mast cell. At the observational end of this spectrum interpretation collapses into description (there is very little difference between interpretation and description), while at the theoretical end of the spectrum there are always two different stories to be told, what is literally observed and a deeper theory-laden interpretation of what is literally observed.

Reflection on our three sample stages in the production of allergic disease establishes a commonsense case for the legitimacy of an observational/theoretical distinction that carries a traditional epistemological force to it. For example, it would seem a tad peevish of someone to insist that the greater theoreticalness of 'immunoglobulin-E molecule' is an entirely made-up theoreticalness, one due to an entirely artificial worry. On the contrary, to put it bluntly, the damn little things are too remote to detect; that is, too remote to detect without a lot of fancy theoretical interpretation of what special machinery churns out for our observation, *no matter what human social system, language-game, or politicized research community you choose.* It is this basic size-niche-based inaccessibility that motivates—has always motivated, in my view—the use of an observational/theoretical distinction by practicing scientists. The distinction is nonarbitrary because what sensory systems we possess given our size-niche as a biological species is nonarbitrary. The theoretical is that which is at a sufficiently greater "distance" from our natural sensory surfaces as to entail a greater risk of error compared to what is much nearer to those same sensory surfaces. Epistemology recapitulates ecology. Admittedly, such a justification of the distinction by example does not resolve its problematic vagueness. But why should we expect the observational/theoretical distinction to be so precise as to be utterly without any vagueness? The philosopher of science Bas van Fraassen has argued that as long as there are clear cases at each end of the spectrum—clear cases of observational terms and clear cases of theoretical terms—we can afford a little waffling and indecision over borderline cases in the middle of the spectrum. The distinction is not rendered useless by the existence of such borderline cases. The swelling and inflammation of the first stage is clearly observable, while the IgE molecules depicted in the third stage are clearly theoretical. It follows that 'urticaria' (the technical term for the kind of swelling in stage one) is an observational term and 'IgE molecule' is a theoretical term. Now what are we to say of the mast cell degranulation that occurs in the second stage? My own preference is to treat 'mast cell degranulation' as a theoretical description, for the theoretical assumptions and preparations involved in producing a photomicrograph of a degranulating mast cell introduce enough risk of error into the situation that we can take the photo to be a rough approximation of mast cell degranulation (in reality, the granules really aren't as shiny as such photos make them out to be, for example). We should be willing, however, to hear evidence that the procedures used to make such a photograph introduce distortions so minimal into the situation that we can pretty much take the photo at face value.

I hope that I have shown the reader that there is a point to the observational/theoretical distinction that goes beyond the accidents of social habits of speech; but that point—the greater risk of error about phenomena remote from our biological size niche—does not make possible a precise and 100 percent clean distinction. Instead, it makes possible an observational/theoretical distinction that, despite its vagueness, has produced a remarkably accurate epistemological track record in science over the last 300 years.

3.2 Correspondence-Rules

Even if a precise criterion were available to us for distinguishing observational terms from theoretical terms, the positivists themselves quickly realized that there are severe problems with the demand that every theoretical term be given an explicit definition in the form of a c-rule. Recall that a c-rule for a given theoretical term *t* attempts to provide purely observational necessary and sufficient conditions for the correct use of *t*. The logical point to saying that the observational side of a c-rule provides necessary and sufficient conditions for the correct use of the theoretical term defined by the c-rule is an important one, for it means any situation in which it is correct to apply or use the term on one side of the c-rule is a situation in which it is correct to apply or use the term on the other side of the c-rule. That is what it means to say that each side of a c-rule refers to a state of affairs *sufficient* for asserting the existence of the state of affairs referred to by the term on the other side of the c-rule. Further, if the term on one side of a c-rule does not apply to a situation, then neither does the term on the other side of the c-rule. That is what it means to say that each side of a c-rule refers to a state of affairs *necessary* for asserting the existence of the state of affairs referred to by the term on the other side of the c-rule. Logicians describe this kind of tight logical connection between two terms by saying that the two terms are coextensive, that is, they refer to properties attributable to *exactly* the same entities (in ordinary English, they are terms true of exactly the same things).

3.2.1 Dispositional Properties

The most influential doctrine about c-rules ever to circulate among practicing scientists themselves (and not just philosophers of science) was **operationalism**. Operationalism was first introduced to a wide scientific audience by the physicist Percy Bridgman in 1927, just as logical positivism was starting up in central Europe. Operationalism did not last long in the physical sciences; but, for reasons that continue to puzzle philosophers of science, it survives to this day with considerable influence in the social and behavioral sciences (especially psychology), where the methodological war cry to "operationalize your variables!" persists among practitioners in certain quarters despite the problems with operationalism that we are about to investigate.

The basic idea of operationalism is that the observational side of a c-rule should specify a purely observable "operation" that must be performed on the system or entity under study to detect, measure, or manipulate the object, property, or event referred to by the term on the theoretical side of the c-rule. To take an example from the behavioral sciences—the current center of the operationalist camp—consider the psychiatric term 'paranoid disorder', which refers to a recognized disease under the DSM-IV (the *Diagnostic and Statistical Manual of Mental Disorders*, fourth edition, the catalogue of psychiatric diseases officially recognized by American medical science and the American medical insurance industry). Clearly, 'paranoid disorder' is a theoretical term, for it refers to a theoretically remote—because "inside" the patient—collection of properties that may from time to time produce observable manifestations in the paranoid patient's behavior. The observable symptoms are distinguishable from the underlying disease process that causally produces them. An operationalist c-rule for 'paranoid disorder', which we can symbolize logically as 'P_d', might look something like this:

$$(\forall x)(P_d x \equiv x \text{ scores in range r on the MMPI}).$$

The idea is that the psychiatric practitioner cannot be sure that she is using 'paranoid disorder' correctly, as applied to a given patient, unless she has administered the MMPI (a standardized psychiatric test) to the patient in question and that patient has scored in range r, which on the MMPI indicates tendencies toward thought and behavior psychiatric theory classifies as paranoid. Such a c-rule is interpreted by the doctrinaire operationalist as being what specifies the scientific meaningfulness of the theoretical term it defines, in this case, 'paranoid disorder'.

Let us turn next to an immunological example. Imagine a misguided allergist who has decided to be an operationalist. Perhaps our allergist is sick of hearing all the skeptics—of which there are more than the reader might think—who claim that "allergies are all in your head," or that "allergies are all psychosomatic." Our allergist wishes to silence the skeptics by operationalizing a host of theoretical terms and descriptions that occur in allergic medicine, terms like, for example, 'allergic to grass pollen'. Let us symbolize this theoretical term in the formalized version of allergic medicine as '$H_1{}^{gp}$' (for 'hypersensitivity [type I] to grass pollen'). As we saw in chapter 1, this term presumably refers to a complex biochemical state of an individual's tissues and serum, a state of affairs that "extends" all the way from the initial genes coding for the production of B cells that produce the IgE antibody molecules specific for grass pollen grains up to the level of mast cell degranulation and the release of a host of chemical mediators of inflammation. But to the skeptics that is all merely a quaint story about tiny things, a story that may or may not be an accurate account of allergic phenomena. Our misguided allergist seeks to shut the skeptics up with a c-rule which will provide the observational cash value of 'H_1^{gp}'. Here it is.

$(\forall x)(H_1^{gp}x \equiv$ a standard skin prick test with grass pollen when given to x produces a wheal within twenty minutes)

A wheal is a perfectly observable raised red bump on the skin where the skin prick test was performed. It would look, in the case of a severe allergy, much like the inflammatory wheal in Figure 3.1.

There are two main problems with operationalist c-rules: one problem that does not really trouble operationalists, and one problem whose solution defeats the whole point of operationalism.

We will deal with the less devastating problem first. If the observable operation specified in a c-rule *defines* a *specific* theoretical term, which accordingly refers to a *specific* theoretical phenomenon, then *different* observable operations performed to detect, measure, or manipulate the *same* theoretical phenomenon actually detect, measure, or manipulate *different* theoretical phenomena. For example, if we use a skin prick test to detect an allergy to grass pollen, and then we use a RAST (radioallergosorbent test)—which is different from a skin prick test—to detect the allergy to grass pollen, then we have not actually detected the same theoretical state of the patient's immune system. Rather, if we are being strictly correct operationalists, we will have to say that the skin prick test is an observable operation that detects the theoretical property of *being-allergic-to-grass-pollen-as-detected-by-a-skin-prick-test*, and that the RAST (which is sufficiently complicated and different from the skin prick test that I will spare the reader its details) is an observable operation which detects the *different* theoretical property of *being-allergic-to-grass-pollen-as-detected-by-a-RAST*. This is at first glance a counterintuitive and unhappy result. But it is a result that most operationalists do not fear much; for there is nothing that makes it impossible that the same theoretical phenomenon could be detected, measured, or manipulated by different observable operations. You simply need some commonsense grounds for supposing that the differences in the operations do not distort or alter the nature of the theoretical phenomenon that the operations detect, measure, or manipulate. It is perfectly possible—indeed, it happens all the time—that linguistic expressions with different *meanings* can nevertheless *refer* to the same thing. For example, there seem to be no believable grounds for holding that the differences between a RAST test and a skin prick test are such as to alter and distort the underlying allergic process in the tested patient. Therefore, it seems reasonable to proceed on the assumption that these two different observable operations detect the same theoretical phenomenon.

The second problem with operationalism is much more serious than the first. In fact, the second problem is a problem that is not unique to operationalist c-rules. It is a problem with the c-rule requirement in general, and therefore it is a problem for *any* positivist model of scientific theories. Surely it is true that a system under scientific study, be it a human allergy patient or the planet Mars, can possess a theoretical property even when no observable operations are being performed on that system—indeed, *even if no observable operation is ever performed on that system.* If this is so, then the observational side of at least some c-rules must be written in the *subjunctive mood.* To see what this would mean, let us rewrite our two previous c-rules in the subjunctive mood.

$(\forall x)(P_d x \equiv$ x *would* score in range r *if* the MMPI *were* administered to x)

$(\forall x)(H_1{}^{gp} x \equiv$ a standard skin prick test *would* produce a wheal within twenty minutes *if* it *were* given to x)

The subjunctive mood ('would', 'were') is necessary because of the interesting but often neglected fact that most scientifically significant theoretical properties are **dispositional properties**; that is, such properties are not always manifested at every moment by the systems that possesses them. The usual simple example used to illustrate dispositional properties is the chemical property of solubility. A substance can be soluble even if it is *never* placed into any solvent during the entire history of the universe. If this is so, then that substance will never observably manifest its solubility to any would-be observers who may be watching for it; yet the theoretical property of being soluble is correctly attributable to the substance.

To return to our own examples, a patient can be allergic to grass pollen even if he or she is never given any kind of clinical operational test for allergy to grass pollen. In fact, that is probably the typical case, for most allergic individuals do not seek medical diagnosis and treatment—they just suffer through.

To be sure, the case of solubility is not quite the same as the cases of having paranoid disorder or being allergic to grass pollen; for a substance can be soluble even if it never dissolves in its entire history, but paranoid behavior and allergic symptoms must occur at least once in a while in an individual if we are to attribute such properties to that person. Yet all three cases *are* similar with respect to the fact that a system can possess any of the three theoretical properties at a particular time without at that same time being subjected to any observable operation specified in any operationalist c-rule you care to construct. In fact, a little thought should convince us that this problem with dispositional properties is not restricted to operationalism. It is a problem for any version of the positivist model, for all versions of the model will require that c-rules be written in the subjunctive mood—as subjunctive conditionals, in fact. For the remainder of this section, therefore, I will discuss the problem of dispositional properties as a general problem for all positivist models of science, operationalist or nonoperationalist.

Let us begin by asking the obvious question: Why is it so terrible for the positivist model if c-rules must be written in the subjunctive mood? There are two answers to that question. First, subjunctive conditionals, conditionals to which we can give the general form 'if A were the case, then B would be the case', are not nearly as precisely definable in mathematical logic as are ordinary non-subjunctive conditionals (for example, 'if A, then B'). This is a technical point which it would not profit us to examine closely. All the reader needs to know here is that the logic of subjunctive conditionals is an unsettled and very unclear area within mathematical logic. This being so, the positivist can no longer claim that using formalized c-rules expressed in the language of first-order mathematical logic provides us with greater clarity and precision in our understanding of the relation between theoretical terms and their "cash-value" in observational terms. But increased clarity and precision were two main virtues that we were promised by the positivist would flow from treating scientific theories as formal languages. That promise turns out to be broken in the case of c-rules.

There is an even worse problem for the positivist that follows from the need to write c-rules in the subjunctive mood: *It lets in everything.* What this means and why it is a big problem for the positivist can be made clear with a little effort. As we saw above, one central methodological and epistemological purpose of requiring c-rules was that it supposedly keeps science honest by ruling out as scientifically significant all bogus theoretical terms for which necessary and sufficient conditions defined in purely observational terms cannot be given. But if one is allowed to use the subjunctive mood in writing c-rules, the corner pseudoscientist can be ushered into the hall of science through the back door. To introduce bogus theoretical terms into a scientific theory the pseudoscientist need only be sure to construct the observational side of the c-rule in such a way as to condition subjunctively the observable manifestations of the bogus theoretical entity on extremely unlikely, never-to-occur-in-the-course-of-history, observable conditions. An example will help illustrate the problem. Suppose a pseudoscientific allergist insists that allergies are due to the invasion of the spinal cord by microdemons called willies. We rightfully ask for a c-rule explicitly defining 'willie', and the quack allergist produces this:

$(\forall x)$(x is a willie $\equiv$ were x heated to 2,000 degrees K for 300 years and then cooled to 100 degrees K in 1,000 seconds, then x would be visible as a tiny gray speck in direct sunlight).

The subjunctively conditioned observational cash-value of 'willie' contained in the right side of this c-rule is an observable operation that is technologically infeasible for us to perform, now and perhaps forever—yet all its aspects are observational. Does this mean that 'willie' is as legitimate a theoretical term as 'IgE molecule'? Of course not. But the general strategy for sneaking such bogus theoretical terms into science through the back door should be clear to the reader: Just make sure that the bogus theoretical term refers to a theoretical phenomenon *so* remote from the level of phenomena our unaided senses allow us to observe directly that it is technologically infeasible, or nearly so, for us to bring about the observable test conditions specified in the c-rule for that bogus theoretical term.

3.2.2 Holism

The philosopher of science Carl Hempel was a positivist who eventually came to recognize the serious problems with c-rules we have just now finished reviewing. Hempel's response to these problems was somewhat radical: *get rid of the requirement that every theoretical term in a scientific theory must be provided with an explicit definition in the form of a c-rule.* For that requirement to make sense, Hempel argued, it must be supposed that the terms of a scientific theory are introduced into the theory one at a time, so that their individual meanings can be specified one at a time. But this supposition is false. Typically, the terms—especially the theoretical terms—of a scientific theory are introduced *en masse,* as one interconnected whole, as one large collection of terms that form an interdependent network, no one term of which has any precise meaning in isolation from all the other terms in the network. For example, 'antibody', 'antigen', 'epitope', 'paratope', 'heavy chain', 'F(ab)', 'Fc', 'Fc receptor', and many others too numerous to list form a collective mass of theoretical terms in immunology that immunologists learn as a coherent, interdependent network of terms. You take them all, or you take none of them as is (if you removed one term from the network the resulting loss would ripple through the rest of the network and slightly alter the uses of all the other terms). For example, the term 'antibody' refers to a kind of protein molecule that is capable of binding to an *antigen,* and the binding takes place when the region of the antibody known as the *paratope* binds to the region of the antigen known as the *epitope.* Further, the paratope is the major portion of the *F(ab)* (fragment, antigen-binding), which consists of the paratope plus part of the *heavy chain* region of the antibody. The *Fc* (fragment, constant) portion of the antibody does not bind antigen at all, but it is capable of binding in certain circumstances to a suitable *Fc receptor* molecule. The interconnections could be drawn out even further, but by now this example of a network of immunological concepts should have convinced the reader of Hempel's point about the interdependence of theoretical terms, perhaps even of all the substantive terms in a scientific theory. No one term in a scientific theory can be explicitly defined by other terms in the theory via logical connectives with precisely defined meanings. The postpositivist philosopher of science W. V. O. Quine extended Hempel's point and turned it into a general philosophy of science that, as mentioned previously, is now called holism.

Quine argued that, not only are the meaningful terms in a scientific theory introduced into the theory as participants in interdependent networks, their meanings are derivative in nature. Only large "chunks" of theoretical beliefs have a "meaning" in any rich philosophical sense of the term. We accept or reject as confirmed or disconfirmed (true or false, if you like) not isolated sentences or single claims involving a specific theoretical term, but rather the larger chunk of theoretical beliefs/terms to which the specific theoretical belief or term belongs. For example, the Quinean immunologist cannot refuse to believe that antibodies exist while continuing to accept the rest of standard immunological theory. That would be impossible, for the meanings of many other terms in immunology depend on the term 'antibody' having a meaningful use. Individual theoretical concepts and their terms don't rise or fall over time; rather, entire chunks of theoretical beliefs and terms rise or fall over time. One important result of such a holistic model of scientific theories is that the observational/theoretical distinction ceases to carry the methodological and epistemological weight it did in the positivist model of scientific theories. For the Quinean holist, theoreticalness is a feature spread through the entire theory. Every term has *some* degree of theoreticalness, for every term is meaningfully interconnected to some network of other terms in the theory, some of which are theoretical terms. We will see in the next chapter when we examine Quinean philosophy of science in more detail that even observationality undergoes a drastic reworking in Quine's hands: It no longer carries the traditional presumption that what is observational is what can be more easily known than what is theoretical. In Quine's hands, the observable is the *intersubjective*. 'Rabbit' is an observational term because the conditions for its correct use are something concerning which nearly unanimous agreement can be reached by all users of the term 'rabbit'.

We will look at Quine's views in considerable detail in chapter 4. For our purposes here, what matters is how the move to a more holistic approach was taken over by some philosophers of science who pushed it in a direction that supported relativist and even antirealist models of science. N. R. Hanson was one such philosopher of science in the late 1950s who coined the jargon phrase '*theory-laden*' to describe his view that *there are no purely observational terms at all.* It is important not to misunderstand Hanson, as many have done. Hanson is not saying that no distinction can be made, at least on a conceptual level, between observable phenomena and theoretical phenomena; he is saying that there is such a distinction, but that all scientific terms fall on the same side of that distinction: All scientific terms are theoretical terms. The reason for this, according to Hanson, is that the meaning of any term in a scientific theory is dependent on other beliefs that are sanctioned by, if not the theory the term itself occurs in, then by some other theory or theories held by the practitioner using the term. To oversimplify a bit, what you observe something *as* depends on what else you believe, where some of those other beliefs are themselves highly theoretical ones. Hanson described this dependence by saying that *all scientific terms are theory-laden.* It is important to understand what is at stake in the dispute over whether or not all scientific terms are theory-laden. What is at stake is nothing less than the hallowed (if abused in relativist circles of late) "objectivity" of science. If Hanson is correct, then what is observed is observed only relative to the larger theoretical context within which the observation takes place. Jerry Fodor once ridiculed this view for being a doctrine based on the notion that "literally, anything can be seen *as* anything"—that is, if enough other background theoretical beliefs are altered. Fodor claimed that it is simply false that you can see anything *as* anything, provided that you are willing to change enough of your other background theoretical beliefs.

Hanson was trained in physics and the example he used to illustrate his radical thesis that all scientific terms are theory-laden was taken from the history of astronomy. Hanson asks us to imagine Tycho Brahe, who believed that the sun moves around the earth, and Johannes Kepler, who believed that the earth moves around the sun, watching together the sun rise on a clear morning. Hanson would have us believe that Brahe and Kepler do *not* observe the same event at all! Brahe "sees" a moving object coming up over the horizon while Kepler "sees" a stationary object coming up over the horizon. It is not, argued Hanson, that Brahe and Kepler both see the same thing but interpret it differently, because they hold different background theories relevant to that thing's nature; rather, Brahe and Kepler literally see different "realities."

We can illustrate Hanson's view with a similar thought-experiment inspired by an episode in the history of immunology. For the first fifty years of the twentieth century immunologists debated and disagreed about the origin of antibodies. What triggered the development of antibodies, and which organs in the body produced them? By the 1950s two competing and incompatible theories were dominant in this debate. The **antigen-template theory** held that the precise chemical structure of an antibody was determined completely by the antigen it fit. The antigen served as a physical template—a kind of mold—upon which the antibody was formed by conforming itself to the shape of the antigen. The **clonal selection theory** held that the immune system *naturally* developed in the organism so that it was genetically capable of producing a huge array of distinct, *preformed* antibodies. The antigen merely served to "select" the correct preformed antibody, which happened quite by chance to fit it (in a kind of "happy accident" that struck the defenders of the antigen-template theory as being simply far-fetched). Let us imagine a champion of each theory, say the young Linus Pauling for the antigen-template theory and MacFarlane Burnet for the clonal selection theory (Pauling and Burnet are both Nobel Prize winners—Burnet was a cofounder of the clonal selection theory, and the *young* Pauling defended the antigen-template theory for a brief period early in his career), observing an unfortunate person allergic to bee venom who has a profoundly swollen arm. A Hansonian would claim that the young Pauling and Burnet do not observe the same thing; for the young Pauling observes the causal aftereffects of antibody-producing cells in the patient's immune system which have been "instructed" by the bee venom on how to make antibodies that fit the bee venom breakdown molecules, while Burnet observes the causal aftereffects of the "selection" by the bee venom of preformed antibody-producing cells in the patient's immune system which are genetically committed to making antibodies that just happen to fit bee venom breakdown molecules. Is our Hansonian philosopher of science correct? Surely the answer is no. The young Pauling and Burnet both observe the same thing—the urticaria (swelling) and erythema (redness) of anaphylaxis (immediate and severe allergy). They differ over the causal goings-on buried at a deeper microlevel of the immune system. They disagree on how their mutual observation is to be theoretically interpreted and explained by appeal to causal micromechanisms. It borders on the absurd to insist that wide theoretical disagreement on ultimate causal mechanisms so infects every other belief on immunological matters that the young Pauling and Burnet hold that the two of them do not observe the same medical phenomenon when they observe a patient's swollen arm due to a bee sting. At the very least, Hansonians and other relativists owe us a stronger argument than the one contained in the kind of thought-experiment we have imagined. In general, we would like to know if the holistic turn in philosophy of science initiated by Hempel and Quine gives credibility to—or even *must* logically lead to—the view that all truths of science are relative to the larger theoretical contexts in which they hold, theoretical contexts that themselves cannot be given objective justifications in terms of further evidence from our sensory experiences. In the next chapter we shall examine that issue.

Further Readings

Rudolf Carnap's "Testability and Meaning," *Philosophy of Science 3* (1936): 420–68; and *Philosophy of Science 4* (1937): 1–40 is the classic presentation of the positivist treatment of the observational/theoretical distinction. See also the older Carnap's treatment of the distinction in his *Philosophical Foundations of Physics*, New York, Basic Books, 1966. Hilary Putnam's debunking of the positivist distinction between observational terms and theoretical terms can be found in "What Theories Are Not," in his *Mathematics, Matter and Method: Philosophical Papers, Volume 1*, Cambridge, Cambridge University Press, 1975, pp. 215–27. Jerry Fodor defends his version of the distinction in "Observation Reconsidered," *Philosophy of Science 51* (1984): 23–43. An interesting treatment of the distinction somewhat related to my sensory surface distance criterion can be found in A. O'Hear, *An Introduction to the Philosophy of Science*, Oxford, Oxford University Press, 1989. The subject of correspondence-rules and dispositional properties was given its classic formulation in Carnap's "Testability and Meaning," previously cited. An extremely lucid but rather technical criticism of the positivist model can be found in Carl Hempel, "The Theoretician's Dilemma: A Study in the Logic of Theory Construction," in his *Aspects of Scientific Explanation*, New York, Free Press, 1965, pp. 173–226.

4

The Underdetermination of Theory

There is a popular misconception among some people about how scientific knowledge is pieced together as a result of experimentation and research. This misconception dates back to the time of Francis Bacon in the seventeenth century, and it is a view of science sometimes called the **naive Baconian** view of science. Now any time that philosophers use the adjective 'naive' to describe some person's view of something, you can bet the house that it is meant as an insult to that person. Naivete is a cardinal sin within philosophy, and what philosophers of science intend to suggest by calling the misconception in question a naive one is that it is, at a minimum, too simplistic and crude to be trustworthy, and, at worst, dangerously misleading and just plain false.

The naive Baconian philosopher views science as the straightforward gathering of "facts" and their systematic organization into useful collections of facts. The idea is that nature will tell us what the facts are, provided we learn how to listen to nature properly. We perform our experiments, and nature rules on the outcomes. We simply have to be sharp enough to devise the right sorts of experiments to distinguish the true facts from the false conjectures. A central presumption of the naive Baconian view is that "crucial" experimentation is the hallmark of science. A crucial experiment would be an experiment that unambiguously favored one or the other of two competing hypotheses under test, no matter the outcome. Suppose an immunologist is wondering whether there are white blood cells in the mammalian immune system whose function it is to identify and kill cancer cells. Two hypotheses need to have the verdict of nature delivered unto them: hypothesis number one asserts the existence of such cells, and hypothesis number two asserts that such cells are not to be found in the immune systems of mammals. A crucial experiment, in this case, would be an experiment whose outcome, no matter what, said yes to one of the hypotheses and no to the other.

It can hardly be doubted that *if* we could design crucial experiments, then we would certainly welcome and use them. The naive Baconian philosopher of science assumes that we can indeed design such experiments, and that, at least in some past instances, science has involved crucial experiments. Not all philosophers of science, however, are naive Baconians (in fact, almost none of them are, anymore). Some philosophers deny that crucial experiments are possible in certain domains of scientific inquiry; other philosophers of science deny that crucial experiments are possible in any domain of scientific inquiry. Pierre Duhem, the French physicist and philosopher of science, argued around the turn of the twentieth century that crucial experiments are not possible in physics. The philosopher W. V. O. Quine, at least during one period of his distinguished career, took the radical view that crucial experiments are forthcoming in no domain of scientific inquiry; indeed, Quine in his more radical moments can be viewed as maintaining that crucial experimentation is a fairy tale of logic nowhere realizable no matter how smart human inquirers are. It is standard practice in philosophy of science to coalesce these two views into one thesis about the epistemological limitations of scientific inquiry, a thesis called the **Quine-Duhem Thesis**.

4.1 The Quine-Duhem Thesis

The Quine-Duhem Thesis has been the source of much debate and controversy. Some philosophers think that it would have been extremely important had it been true, but that it is in fact false; others think that it is extremely important and true. Still other philosophers think it is true but trivially so; and a small group think that, not only is it false, but it would have been trivial had it been true. One problem with the debate as it has been carried on up to the present has been the lack of an agreed on rendering of the Quine-Duhem Thesis. Different philosophers give slightly different versions of it, and sometimes the differences are enough to justify very different methodological, epistemological, and ontological implications. Quine himself has not helped matters out. Over the years he has, to be polite about it, waffled a bit on just what he intended the thesis half-named after him to entail. Some philosophers see in this waffling an outright retreat by Quine from his original radical position. I prefer to interpret Quine's waffling as a form of healthy refinement. He came to believe that he'd presented the thesis at first in a misleading way, and he later wished to refine and amend the thesis so as to expose better its plausibility and claim to accuracy. What the Quine-Duhem Thesis claims is this:

> *Any* seemingly disconfirming observational evidence can *always* be accommodated to *any* theory.

In other words, the Quine-Duhem Thesis denies that the disconfirmation of a theory can be forced upon a practitioner by the evidence itself. Nature does not rule, we do; we do in the sense that our rejection of the theory under test on the basis of evidence is an inferential decision made by us, not by nature. For, if we really wanted to, if we were willing to change enough of our other beliefs, we could render the evidence in question consistent with the theory under test. How is this possible, you ask? It is easy to see how it could be done once we move to a holistic conception of scientific theories. Recall that according to the holistic conception of theories the terms and laws of a theory form one interdependent network. An important and obvious consequence of this is that it is always a large chunk of a theory that faces an experimental test, never one isolated prediction. It follows that a piece of seemingly disconfirming observational evidence never contradicts a single isolated claim, it contradicts the large chunk of theory that the claim that otherwise appears to be directly under test is part of (Quine in some of his earlier and more radical moods argued that it contradicts the *entire belief system* of the practitioner, but he later backed away from such an extreme version of holism). Obviously, therefore, we could "save" the claim that the context might seem to suggest is directly under test by adjusting other claims in the large chunk of theory with which it is connected. There is more than one way to accommodate seemingly disconfirming evidence besides throwing out the claim that appears to be directly under test. This is the central insight of the Quine-Duhem Thesis, an insight that would not have been reached had there not been a move to a more holistic model of scientific theories.

It is important not to misunderstand the Quine-Duhem Thesis in such a way as to make it appear absurd. Some philosophers, prominent among them Larry Laudan, take the thesis to entail that there are no defensible, rational grounds for rejecting one theory on behalf of another where the competing theories are all consistent with the observational data. Quine nowhere proclaimed such a brutal methodological quandary. His point was a logical one; or, put another way, his point was a metaphysical one: How theories are adjusted in the light of contrary observational evidence is a human affair, a decision that we, and not nature, make. So, as a matter of mere logical consistency, it would always be possible to render consistent any theory with any seemingly disconfirming evidence by making enough adjustments in other beliefs. Quine was concerned to insist on this as an expression of his deep commitment to **pragmatism**. How we adjust theories to accommodate contrary observational evidence is a highly pragmatic affair: We adjust so as to maximize fertility of the resulting theory in generating new predictions, to maximize consistency with as much of our former system of beliefs as possible, to maximize the overall simplicity of the resulting theory, and to maximize the theory's modesty so that its claims are as narrowly drawn as possible. As a pragmatist Quine believes that such pragmatic features "track the truth" in the sense that a maximally fertile, familiar, simple, and modest theory is, in virtue of possessing those traits, that much more likely to be a true theory. This faith that pragmatic features of a theory track the truth is, of course, a contentious claim to nonpragmatists. For our purposes what matters is that this process of adjustment is in our hands, not nature's—and that was the deep point buried in Quine's rather polemical presentation of the thesis.

To proclaim such a thesis is an act of bold conjecturing; to find illustrations of it in the actual history of science would be much more impressive. The usual cases used to illustrate the Quine-Duhem thesis are taken from the history of physics. This was no accident. Physics does have a prima facie claim to being a fundamental science in virtue of its universal scope and maximum generality, so it should not be surprising that any special methodological quandaries somehow "necessary" to human science would show up in physics first and in especially nasty garb. In fact, Duhem thought—mistakenly, in my view—that the thesis with which his name subsequently became associated was a methodological oddity unique to the purely physical sciences. He claimed that it did not occur in the life sciences, for example. (Quine was wiser on this point because he realized that the methodological problem, if it was a problem, would be general across all domains of scientific inquiry.) One standard case illustrating the thesis involved the fall of Newtonian physics and the rise of Einsteinian physics. It was argued that the observational evidence for the relativity of space and time did not logically falsify Newtonian absolute space and time, for absolute space and time could be saved by postulating the existence of bizarre "universal forces" that bend light rays, shrink measuring rods, and slow down clocks, forces which by their intrinsic nature are forever and ever undetectable (because they affect any experimental device designed to detect them in such a way as to remain undetectable). Here was a case of two incompatible theories each consistent with all actual *and possible* observational evidence. The physicists coined a phrase to describe such a situation, saying that in such situations theory is *under*determined by observational evidence, meaning that the evidence cannot by itself determine that some one of the host of competing theories is the correct one. This was sloganized into the phrase that serves as the title of this chapter: the underdetermination of theory. The important impact of the underdetermination of theory was alleged to be that the subsequent choice of which theory to commit our resources to could not be a choice made solely on the basis of the evidence of nature, but it also had to involve the use of special methodological principles that must be adopted to break the competitive deadlock between the evidentially equivalent competing theories. Quine initially proposed the pragmatic virtues mentioned above as being the most truth-maximizing group of methodological principles that could be used to choose among evidentially equivalent competing theories.

The underdetermination of theory threatens a couple of pet philosophical positions. It would seem to threaten scientific realism, for it suggests that even in the **limit of scientific inquiry** (a fictional point in the future when science has evolved as close to a completed form as it can get without actually being perfected) there may be no single theory that has survived the competition as the correct one but rather a host of mutually inconsistent but equally adequate (to the observational evidence) theories. This is a serious worry, and its seriousness and complexity require that we give the entire topic of scientific realism its own chapter. I will hold my remarks until chapter 10. For now we should note that the underdetermination of theory would seem also to threaten even more directly the so-called hypothetical-deductive method, which Karl Popper enshrined as the hallmark of mature science under the name of **falsificationism**. This is an equally serious charge, and to its consideration we now turn.

4.2 Popperian Falsificationism

Many science textbooks begin with a chapter which provides a superficial treatment of "scientific method," and the majority of those chapters adopt some more or less falsificationist account of scientific epistemology (although few of them use the term 'falsificationism', instead preferring 'hypothetico-deductive method'). Popperian falsificationism conceives of scientific practice as proceeding under the rules of logic known as *modus ponens* and *modus tollens*. First a theory is boldly conjectured—free-form, as it were—and then an observational prediction is deduced. An experiment is then designed with the express purpose of showing that the prediction will *not* be fulfilled. If that happens, the theory from which the prediction was deduced must be wrong. On the other hand, if the prediction is borne out by the experiment, then the theory from which the prediction was deduced is not thereby confirmed, but merely corroborated (that is, it has survived one attempt to falsify it). This would be a happy tale if only nature cooperated in the right way. Unfortunately, the Quinean philosopher of science wishes to spoil the happy ending of the tale.

Let us construct an example to illustrate how the underdetermination of theory and the Quine-Duhem Thesis which generalizes such underdetermination threaten Popperian falsificationism. Suppose we wish to test the *immune surveillance theory of cancer*, a theory under which the immune systems of mammals contain white blood cells that kill cancer cells naturally. The immunologist means by 'naturally' that these alleged white blood cells can differentiate cancer cells from normal cells on first contact with them, that a prior "sensitization" through prior exposure to the cancer cells is not required for the recognition or killing to take place. Indeed, in the 1970s immunologists claimed to have discovered such white blood cells in the immune systems of mammals, cells that they named NK (natural killer) cells. Do the immune systems of mammals contain NK cells, or not? What a perfectly fine puzzle for a practitioner of immunology to spend a couple of years working on. There are a number of standard procedures for testing such a claim. One standard procedure is called a ^{51}chromium release assay. It is possible to describe this procedure in a superficial way so that it sounds pretty straightforward. We incubate target cancer cells in solution with radioactive chromium. Some of the cancer cells take up the chromium into their cytoplasm where it becomes bound to proteins. We then wash away any excess free radioactive chromium, plate out the cancer cells into little wells, and add to each well white blood cells extracted by centrifugation from the blood of a subject mammal. The NK cells in the blood serum added to each well should kill some of the cancer cells and thereby release radioactive chromium into the supernatant, the remaining liquid in the well. After a suitable incubation period, we therefore check the supernatant of each well for radioactivity using a spectrometer. The degree of radioactivity is correlated with the degree of killing power of the NK cells.

But things are not as simple as this story might suggest. The story is a heavily edited version of what actually goes on in a ^{51}chromium release assay. It is almost misleading in how it oversimplifies what is in reality a remarkably convoluted procedure. It is misleading in that it encourages the idea that the single hypothesis that there are NK cells in the mammalian immune system is facing a test by itself. This could hardly be further from the truth; for the procedure in question makes use of a massive chunk of interdependent laws, theories, and claims. A ^{51}chromium release assay is downright ingenious in how it hooks together various experimental processes, let alone a wonderful example of the holistic interdependence of theoretical terms, laws, and claims; so we have good cause to examine the procedure in some detail.

The first thing we need to understand is how complicated this procedure is bound to be. First, we shall need some way of isolating the alleged NK cells from all the many other components found in the mammalian immune system. Second, we will need a way of isolating targets, that is, cancer cells, for the NK cells to kill. Third, we shall need some way of marking or labeling the targets so as to be able to tell if they have been killed by the NK cells. These facts alone ought to make us skeptical about the prospect of singling out for testing the lone claim that there are NK cells in the mammalian immune system. There is bound to be too much interconnection with other theoretically characterized processes to allow us to be able to isolate a lone hypothesis in that way. The testing itself forces us to expand the range of what is being tested. To use the fashionable terminology for this situation, we must use a number of *auxiliary* theories, laws, and beliefs to test the original hypothesis in which we were interested. The upshot is that we simply cannot test the original hypothesis in isolation from the rest of our theories and beliefs. Let us see why not in some detail. I shall present one standard way of performing a ^{51}chromium release assay. I shall break the procedure into steps and comment on each step where appropriate. The procedure is adapted from one described by Leslie Hudson and Frank Hay. Figure 4.1 provides an oversimple (as we shall shortly see) representation of this procedure in which the actual details are as follows.

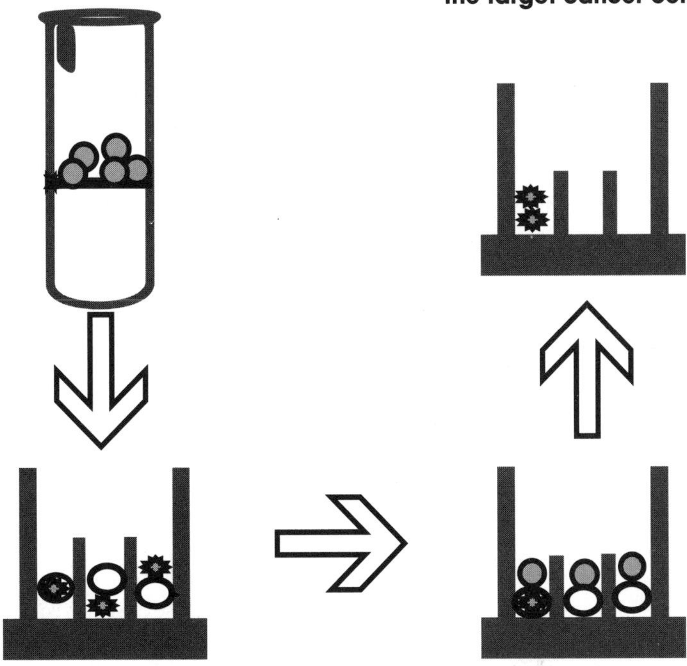

Figure 4.1. A ^{51}chromium release assay.

1. Draw blood from the subject mammal (perhaps a mouse or even a human being), and isolate a *class* of white blood cells called mononuclear cells by density-gradient centrifugation.
   Comment: Current theory claims that NK cells would be members of the class of mononuclear cells. Further, the drawn blood must be heparinized to prevent its clotting in the test tube. Density-gradient centrifugation is a standardized procedure in which different classes of white blood cells are

separated from each other by centrifugation (whirling at great speed) in a tube packed with Ficoll.

2. Wash the mononuclear cells three times in tissue culture medium by centrifugation at 150 g for 10 minutes at room temperature.
   Comment: This is a step that purifies the mononuclear cells by washing out impurities.
3. Cut the washed mononuclear cells to about 5,000,000 per milliliter by dilution with tissue culture medium containing 5 percent fetal cow serum.
   Comment: The killing power of NK cells that we are seeking to demonstrate may be dependent on the number of NK cells per cancer cell, so we must have a rough estimate of cell numbers throughout the entire procedure.
4. Chose a target cell line. Let us use the leukemia cell line designated K-562. This cancer cell line we must now label somehow. Let us label the K-562 cells with a radioactive form of chromium by mixing 1,000,000 of the cells with 370,000 becquerals of radioactive chromate $^{51}CrO_4$. After mixing we incubate the mixture at 37 degrees Celsius in a water bath for 90 minutes, stirring the mixture every 30 minutes.
   Comment: Here a host of complicated theoretical commitments enter the picture. One troubling fact about NK cells is that most cancer cell lines available for laboratory use are resistant to them. K-562 happens to be a cancer cell line that is not resistant to them. If we used a different cell line, say, the bladder cancer cell line designated T-24, which is resistant to NK cells, we would likely have guaranteed ourselves a negative outcome to the procedure. Further, a whole host of assumptions must be made about how the cancer cells will take up the radioactive chromium in the course of normal metabolism. If the cells did not absorb the chromium into their cytoplasm where it binds with proteins, the whole procedure would be useless. Further, we must assume that the radioactivity does not structurally or biochemically degrade the cells so much that the NK cells cannot recognize them as cancer cells. In fact, we must assume that the radioactivity does not kill all the cancer cells outright; otherwise, there would be no viable cancer cells that could serve as targets for the NK cells. Step 6 together with the spontaneous cell death rate after labeling (which would be available as a laboratory standard value) help to lessen these last two worries.
5. Wash the incubated target cells three times by centrifugation, then count the number of cells per milliliter in a hemocytometer, separating different concentrations according to the protocol in step 7.
   Comment: It will be necessary to mix the mononuclear cells with the target cancer cells in a series of different ratios, for the killing power of NK cells varies with the concentration of cancer cells in their environment. Hence,

we must mix a constant number of mononuclear cells per milliliter with increasing concentrations of cancer cells per milliliter. A hemocytometer is a device for counting the number of viable cells per unit area. It requires visually counting cells in a microscope by superimposing a grid on the cell suspension. Each higher concentration of cells per milliliter is saved in its own tube.

6. Take 100,000 labeled K-562 cells and kill them by freezing and thawing (3 times at 37 degrees and −20 degrees Celsius); then centrifuge down any insoluble material and count radioactivity using a spectrometer. Use this value as the maximum (100 percent) ^{51}chromium release, which is a number we will need to calculate the killing power of the NK cells in step 11.
   Comment: This step is required so that we may compare how close to 100 percent of the cancer cells the NK cells can kill.
7. Mix 100 microliters of the target cells and 100 microliters of the mononuclear cells together in a series of suspension wells (little holes in a micro-culture plate) according to the concentrations in the protocol presented in Figure 4.2.
8. Centrifuge the microculture plate containing the suspension wells at 50 g for 15 minutes at room temperature.
9. Incubate the wells for 4 hours at 37 degrees Celsius in a humidified incubator gassed with 5 percent CO_2 in air.
10. Remove 100 microliters from each well into tubes; cap and count the radioactive emissions per minute using a spectrometer.
11. Calculate the specific target cell death rate per well as follows: The percent killed equals the observed count per minute minus the spontaneous count per minute divided by the maximum count per minute minus the spontaneous count per minute times 100. In symbols: % K = obs. c.p.m. − spon. c.p.m./max. c.p.m. − spon. c.p.m. × 100
    Comment: The spontaneous count per minute would be available to us as a standard laboratory value—at least it would be so available if our immunology lab did a lot of cellular assays involving radiolabeled cells suspended in tissue culture, so I have omitted the details of how it would be calculated in the above specific assay. Those further details would complicate the assay even more.

	Well 1	Well 2	Well 3	Well 4
Millions of mononuclear cells per milliliter	5	5	5	5
Millions of target cells per milliliter	30	60	125	250
Mononuclear cell to target cell ratio	1 to 6	1 to 12	1 to 25	1 to 50

Figure 4.2. A sample protocol for a ^{51}chromium release assay.

12. Calculate the mean standard error for each trial run.

    Comment: This step is simply sound experimental practice. Any practitioner who is adequately trained would have the mathematical background to calculate this statistical quantity.

13. If there are NK cells among the mononuclear cells, then they will kill some of the radiolabeled cancer cells and thereby release the radioactive chromium into the supernatant. Hence, we should get a percent killed value greater than zero for at least one of the wells.

Who else but a trained immunologist would have thought that the simple story we began with actually hid the kind of interdependent complexity we have just finished working our way through? Popperian falsificationism was too naive a picture of actual scientific practice. Falsificationism would have us believe that a general theory like the immune surveillance theory of cancer is straightforwardly testable. We simply note that it deductively implies a certain observable result—in this case, that we should observe a count per minute rate above the spontaneous rate in at least one of the wells—and then arrange for the right sort of experiment in the attempt to falsify that prediction. But our detour through the reality of the experimental laboratory has given the lie to that simple account of experimental method in science. The immune surveillance theory of cancer does not by itself imply any observational prediction, and even Popper admitted as much about any general theoretical hypothesis (that is, he admitted that no general hypothesis by itself implies anything about an actual particular state of affairs). What implies an observational prediction is that theory together with a myriad of interdependent beliefs, presumptions, guesses, and other theories. So the *modus ponens* part of the Popperian model of scientific method involves a complex conditional. It is not that

> theory T implies observational prediction O,

but rather, it is that

> theory T plus theory R plus theory M plus belief B plus assumption S plus assumption Y plus hunch C plus ... imply observational prediction O.

But if this is the case, then the *modus tollens* part of the Popperian model of scientific method is logically undermined. From the falsity of observational prediction O we can no longer automatically infer that it is theory T that is false. All we may legitimately infer, *as a matter of logic alone*, is that at least one of T, R, M, B, S, Y, C, ... is false, and that is just what the underdetermination of theory by evidence is all about, and it is also just why the Quine-Duhem Thesis is a possible epistemological quandary for human scientists. In our example, if the tested wells do not turn up with a percent killed greater than zero, it would surely be hasty—if not flat out reckless—to infer that there are no NK cells in the immune system of the mammal whose blood was used in step 1, and therefore that the immune surveillance theory of cancer is incorrect. Instead, we would immediately point to the many places in the long procedure where error could have been introduced into the process, either experimental error because we performed a step improperly, or else theoretical error because the theoretical assumptions that justified a particular step or part of a step were not correct.

A sensitive appreciation for the epistemological impact of the underdetermination of theory is essentially what killed Popperian falsificationism as a plausible model of science. The recognition, as our own example shows, that holistic underdetermination extends to domains of inquiry other than the physical sciences gave the lie to Duhem's attempt to limit underdetermination of theory to physics. But there are philosophers of science who, without being defenders of Popperian falsificationism, nevertheless take a debunking attitude toward the underdetermination of theory and the Quine-Duhem Thesis. They believe that too much respect has been accorded both the thesis and underdetermination. These philosophers of science believe that the extent of underdetermination has been exaggerated and that it is mostly a myth. And because you cannot erect an accurate methodological thesis on a myth, the Quine-Duhem Thesis turns out to be so much armchair worrying about nothing. It will surely be of importance to see if these philosophers of science are correct.

4.3 Is Underdetermination a Myth?

The Quine-Duhem Thesis has drawn much commentary from thinkers spread across the scientific and intellectual spectrum: philosophers of science, sociologists of knowledge, postmodernist cultural critics, social constructivists, and many others. Some thinkers have sought to draw certain implications from the thesis that are of a rather shocking kind; for example, that there is no such thing as scientific truth independent of contingent scientific practices, or that science has no more call to be the arbiter of truth about any matter than does The New-Age Society for Sublime Consciousness, or any other trendy and mystical system of thought. Many attacks on the rationality of science have been mounted in the name of the underdetermination of theory, the idea being that because evidence underdetermines theory choice, theory choice must be a matter of how various nonevidential and "irrational" factors affect the professional (but really personal) decisions of practitioners. We shall examine in depth some of these claims in chapters 8 and 9. There I will argue that the above kinds of radical implications do not follow from a proper appreciation of the Quine-Duhem Thesis and the underdetermination of theory which it presupposes. Accordingly, attempts to use the thesis to support such claims amount to misuing it.

Now from the sole fact that someone misuses a doctrine we may not immediately assume that the doctrine is to blame. Many thinkers have abused the Quine-Duhem Thesis, and they have misunderstood the underdetermination of theory, but that fact does not by itself mean that there is something bogus about the thesis when it is properly construed. Yet some philosophers of science do seem to want to blame the thesis itself for the cognitive sins of those who have misused it. The philosopher of science Larry Laudan harbors a large dose of hostility toward the Quine-Duhem Thesis, apparently because of the encouragement the thesis has provided to the enemies of science found among critics of society and culture who work in the various humanities disciplines. Laudan wishes to debunk the Quine-Duhem Thesis as being no more than a piece of philosophical hyperbole—at least it is no more than that in Quine's hands. If Laudan is correct in his critique, then we have witnessed a lot of methodological handwringing over the underdetermination of theory for no good reason. Further, if the thesis is an exaggeration, then the threatened philosophical positions mentioned earlier may not be as vulnerable as the thesis seems to imply that they are.

What makes Laudan's approach to the underdetermination of theory so interesting is that he distinguishes several distinct kinds of underdetermination. The different kinds have different methodological implications as well as different degrees of plausibility in the light of the history of science. First, we must distinguish a *descriptive* from what Laudan calls a *normative* version of underdetermination. The descriptive version is simply the claim that scientists are psychologically able to hold on to, and have in the past held on to, a theory in the light of seemingly disconfirming evidence by adjusting other beliefs, that this is a bare possibility that has occurred as a matter of historical fact. The normative version of underdetermination claims not merely that this is a possible state of affairs, but that it is justifiable, or rational, for scientists to "save" a theory in the way the thesis says they can save it.

Second, we need to distinguish *deductive* underdetermination from what Laudan calls *ampliative* underdetermination. The deductive version claims that theories are deductively underdetermined by evidence; that is, that the evidence does not deductively entail a single correct theoretical account of the phenomena in question. The ampliative version claims that even if we throw in inductive forms of inference, that is, probabilistic logic, the evidence still underdetermines theory—it is not even the case that the evidence inductively implies a single correct theoretical account of the phenomena in question.

Third, we must distinguish what Laudan calls a *compatibilist* version of underdetermination from an *entailing* version of underdetermination. The compatibilist version claims that theories are underdetermined by evidence in the sense that any theory can be rendered logically consistent with any evidence that otherwise would seem to contradict it (by adjusting suitably elsewhere). The entailing version claims that logical consistency (that is, lack of contradictoriness) is not enough, that evidence underdetermines theory in the sense that any theory can be made so as to entail logically any evidence that otherwise would seem to contradict it.

Finally, we need to distinguish a *nonunique* version of underdetermination from an *egalitarian* version. The nonunique version of underdetermination claims that for any theory and any body of evidence either compatible with or entailed by the theory (depending on which choice of the previous two versions one has made), there will be *at least one* other theory logically incompatible with the first theory which also is either compatible with or logically entails that evidence. The egalitarian version claims something much stronger: that there will be an infinity of such competing theories; in fact, any and every theory is compatible with or entails any evidence base you choose. All theories are epistemologically equal. For ease of presentation let us set out the four choices. Underdetermination can be either:

descriptive or normative

and

deductive or ampliative

and

compatibilist or entailing

and

nonunique or egalitarian.

Laudan saddles his targeted opposition with the strongest set of choices among the above alternatives. He argues that for the science basher to be able to use the Quine-Duhem Thesis to mount the sort of attacks on science he or she wishes to mount, the kind of underdetermination in question must be one that is normative, ampliative, entailing, and egalitarian; that is, he or she must claim that any theory can be adjusted with respect to any evidence so as to entail that evidence, and that it can be rational to do this because not even when ampliative methods are used does evidence rule out even a single competitor theory. We shall see in chapters 8 and 9 below that Laudan is correct only with respect to the last choice: Many contemporary science critics do tend to assume that the Quine-Duhem Thesis implies that all competing theories are epistemologically equal (they adopt egalitarian underdetermination). But there is little evidence that those same contemporary science critics hold to the other three choices Laudan insists that they must adopt. Some social constructivist science critics would reject the first choice outright on the grounds that the descriptive/normative distinction is itself a phony dichotomy. The same science critics also tend to lack the necessary interest in formal logic that would be required even to appreciate the remaining two distinctions between deductive/ampliative underdetermination and compatibilist/entailing underdetermination, distinctions that involve essentially concepts that presuppose formal logic. The worry therefore arises that Laudan is attacking a straw person.

One thing is for sure, though, he is not attacking Quine himself. In his most recent discussion of the Quine-Duhem Thesis, Quine commits himself to a form of underdetermination that is descriptive, compatibilist, and nonunique. Whether a Quinean must adopt deductive or ampliative underdetermination is less clear, as Quine never addressed that issue specifically. What he has written is consistent with either version, but certainly his early presentations of the thesis seemed to assume a deductivist form of holism. I will grant for the sake of argument that the Quinean supporter of underdetermination must adopt an ampliative version of it. Still, my point remains. Laudan has sunk the Quine-Duhem Thesis only by building its hull with large holes in the first place. Laudan admits that it is likely that theories are deductively underdetermined in the sense of nonunique underdetermination. As for compatibilist underdetermination, Laudan argues that it may be true, but underdetermination would be "trivial" if so interpreted because he argues that compatibilist underdetermination takes place by simply removing from the picture any component claims that implied the falsified prediction *without replacing them by other claims*, and so the resulting theory is consistent with the evidence but has surely lost some of its explanatory power as well as some of its predictive power. Further, Laudan argues that nonunique underdetermination is harmless because it cannot support the sorts of global attacks on scientific methodology that critics of science mount against it. It cannot support those attacks because Laudan reads those attacks as boiling down to the claim that science is a "waste of time" in the sense that its methods cannot in the end guarantee a rationally justified choice between rival Theories of Everything.

We shall see in chapters 8 and 9 that Laudan is correct in his latter point: The current crop of science critics tend to adopt an egalitarian conception of underdetermination and proceed to attack science on the ground that its methods cannot generate a unique theoretical account of a given domain of phenomena—its methods can only generate a host of epistemologically equal but mutually inconsistent theoretical accounts of a given domain of phenomena. We will investigate in chapters 8 and 9 whether there is anything to this complaint against the methods of science.

With regard to Laudan's first point about compatibilist underdetermination, that it can only be interpreted trivially and when that is done the resulting theory has paid too high a price for its "salvation," we need only note that the sort of complaint Laudan has about this high price couldn't be more in the spirit of Quine. Quine never said that all adjustments that might save a theory from refutation in the light of seemingly disconfirming evidence are equally preferable adjustments. Some adjustments that save a theory virtually destroy the theory's usefulness. They do so by removing beliefs that were and are needed to explain related phenomena, to make novel predictions in new domains, and so on. That is precisely why Quine said that his form of holism did not lead to an "anything goes" model of scientific practice. Quine is not an epistemological anarchist. Quine held that we apply pragmatic principles of choice to pare down competing theories, principles like those already mentioned in section 4.1: simplicity, fertility, modesty, and conservatism. These principles help us to reject adjustments that salvage a theory at too high a cost in the sense of being salvage operations that undermine the theory's prior explanatory and predictive power. If Laudan holds that this is a trivial result, then he has obviously never encountered a science enthusiast who believes that nature delivers a single, certain, and clear verdict on our inquiry, if only our inquiry were ingenious and careful enough, and carried out to some final and mature form.

My argument suggests that Laudan has erected for his target a version of the Quine-Duhem Thesis that is so extreme that few people could be found who hold to it, least of all a Quinean supporter of the underdetermination of theory. Quine himself has the better attitude. He is not sure if the methods of science will determine a unique Theory of Everything in the limit of inquiry. He surmises that they may not, and he argues that it is at least possible that they may not. What he denies is that we have a guarantee beforehand that they will. We shall just have to wait and see, which is to say that some of our progeny will get pretty darn close to seeing what we can only wonder and argue about. As a pragmatist Quine thinks that answering the question of whether our scientific theories accurately represent an inquiry-independent reality can be postponed anyway. What matters in the present is that we have a canon of methods, an epistemology of daily practice, which underwrites enormous confidence in the claims of mature scientific theories. Let us see next why Quine thinks that those who believe that science truthfully represents an inquiry-independent reality have good cause to be encouraged even in the face of the Quine-Duhem Thesis.

4.4 Pragmatism and Realism

Are there NK cells in the immune systems of mammals? Pragmatism is a philosophical view that would answer this question with a hypothetical statement: There are NK cells in the immune systems of mammals if positing the existence of such cells maximally simplifies, renders maximally fertile, conservative, and modest, the sensory experiences (and our theoretical reflections on them) which we undergo in our immunological inquiries, as compared to what would result on those same criteria if we did not posit their existence. In other words, there are NK cells if the NK-cell hypothesis "works." Nature, after all, pushes back against our theorizing, and if a hypothesis fails to do the work for us that we want it to do, that is a rational ground for becoming suspicious that the hypothesis might be barking up the wrong tree, ontologically speaking. That allergies are caused by the willies invading the spinal cord is a hypothesis that does not work: When we perform the willies-exorcism rite, the allergy patient fails to get better; a williesometer built with a $75,000 grant from the National Association for the Investigation of Willies fails to detect any willies emanating from the backs of allergy patients. If these sorts of failures are widespread enough, if they go on long enough, if competing explanations of allergic phenomena appear on the scene that seem to work better, then not only are we rational in abandoning the willies theory of allergies, but there are no willies around for us to abandon theoretically anyway.

The word 'posit' is a Quinean term of art. Quine pioneered the use of this word as a noun rather than a verb. As a verb it is roughly equivalent in meaning to 'postulate'. Taken as a noun one can see the general idea: The existence of electrons, for example, is suggested by the success we have in doing physics with an initial postulation that there are electrons. Such overwhelming and consistent success is the only evidence there could be that there are such things as electrons, that they exist as inquiry-independent entities that push back against our theorizing. Posits which underwrite sensory success "earn" the status of being "realities." But Quine is insistent that a posit is not necessarily after-the-fact. It is not necessarily made in order to help organize our prior sensory experience; rather, it can be needed to make that experience coherently understandable in the first place. Reality, argues Quine, is discovered rather than invented, but its discovery is built up over time via inquiry-sensitive posits (some of which work and therefore count as discoveries in the end, some of which don't work and are eventually discarded).

Such is the pragmatist's attempt to make an end run around the thorny issue of whether science is about an inquiry-independent reality. "It all comes out in the doing," the pragmatist claims. We learn by doing, we know by doing, we believe by doing, so only by doing will these "metaphysical" questions be answered. What we must resist is the attempt to freeze knowledge in its current imperfect state while we try to read off of it what the ultimate Theory of Everything will contain. But a critic will wish of course to challenge the pragmatist's seemingly simplistic equation of what works with what is the case. Suppose some theory T works in all the senses that a pragmatist like Quine thinks count. Why should that be taken as a ground for thinking T is a true representation of some inquiry-independent reality? Why assume that workable theories track the objective truth? Quine's answer is a slight duck of the issue. He misconstrues this complaint as the charge that

> if a posit is based on pragmatic usefulness, then it is a posit of an unreal entity.

This conditional is false, according to Quine, because its contrapositive

> if a posit is of a real entity, then the posit is not based on pragmatic usefulness

is *obviously* false (for how could the real be the useless?). I don't know how the real could be the useless, but that isn't the point that the critic of pragmatism is depending on anyway. The critic is not claiming that being a posit based on pragmatic usefulness is sufficient for being a posit of an unreal entity. The critic is claiming that workable theories don't have to work because they track the truth: They could do good work for us and still be false.

At last we have reached the ontological nut in the center of the huge shell of epistemology we've been hacking our way through since this chapter began. Why should it be the case that what works is what's true, that what works reveals what is real? Well, argues the pragmatist, what other explanation for a theory's working well over a long period of time could there be? What other explanation for why practitioners achieve their goals using the theory could there be? What else could being real consist in but that we can successfully work with it? Clearly, this dispute depends heavily on the sense of 'work' each side is presuming to be at issue. Exactly what kind of work is it that we ask our scientific theories to do for us? And how can we measure or judge when they have done such work well or poorly? Before we can settle this ontological dispute, we shall need a fuller understanding of the tasks and chores we expect theories to carry out for us and the means we use to ascertain if, when, and to what degree they have or have not succeeded in carrying them out. We begin constructing that fuller understanding in the next chapter.

Further Readings

Pierre Duhem's initial presentation of the underdetermination of theory by observational data is in *The Aim and Structure of Physical Theory*, Princeton, Princeton University Press, 1982. W.V.O. Quine's classic account of holism was first presented in the closing pages of "Two Dogmas of Empiricism," in his *From a Logical Point of View*, 2nd. ed., New York, Harper, 1961, pp. 20–46. Quine expanded his holistic model in chapter one of *Word & Object*, Cambridge, Mass., MIT Press, 1960. A more popularized treatment of holism may be found in W.V.O. Quine and J. Ullian, *The Web of Belief*, New York, Random House, 1970. Karl Popper proposed his falsificationist model of science in *The Logic of Scientific Discovery*, New York, Basic Books, 1959. See also Popper's "Truth, Rationality, and the Growth of Scientific Knowledge," in his *Conjectures and Refutations: The Growth of Scientific Knowledge*, New York, Harper, 1965, pp. 215–50. Larry Laudan attempts a general debunking of the Quine-Duhem Thesis in "Demystifying Underdetermination," *Minnesota Studies in the Philosophy of Science, Volume XIV*, edited by C. Savage, Minneapolis, University of Minnesota Press, 1990, pp. 267–97. Quine confronts the ontological implications of holism in "Posits & Reality" in his collection of papers *The Ways of Paradox*, rev. ed., Cambridge, Mass., Harvard University Press, 1976, pp. 246–54. For Quine's most recent pass at these issues see sections 6 and 41 of his *Pursuit of Truth*, Cambridge, Mass., Harvard University Press, 1990. The deadly technical procedure for doing a ^{51}chromium release assay was noted in the text, but the reader wishing to obtain an even greater sense of the complexity of laboratory practice in immunological science should consult the remaining pages of L. Hudson and F. Hay, *Practical Immunology*, 3rd. ed., Oxford, Blackwell Scientific Publications, 1989.

5

Reductionism, Antireductionism, and Supervenience

The beginnings of science can be traced historically to the ancient city of Miletus, which was located on what is now the western coast of Turkey. Miletus was a commercial seaport in a region known to the locals of the time as Ionia. Around 600 B.C.E. a great sage lived in Miletus whose name was Thales (THEY-lees). Among other achievements attributed to this wise person was the prediction of an eclipse of the sun that took place in the year 585 B.C.E. (scholars believe that Thales predicted it to the year, not to the day—a still impressive degree of precision given the primitive state of organized astronomy at the time). Thales was the first scientist. He is traditionally considered the first philosopher as well. Why was Thales the *first* scientist? Because he tried to provide an account of the universe in entirely "natural" terms. He is said by Aristotle to have claimed that

> the *arche* (first principle) of all things is water,

meaning that all things are made of water, that all things *are* water. This was something brand new at the time, for what was then the traditional manner of explaining most things involved making appeals to supernatural deities and magical forces of various kinds. Far from being a silly hypothesis, Thales' theory brought theoretical speculation about the universe down from the supernatural realm and into the world of ordinary human experience. Some philosophers argue that Thales had rational grounds for choosing water as the basic "stuff" of the world. Water was in ancient times the only natural substance known to exist in all three states of matter—gas, liquid, and solid. Further, all forms of life that the ancients knew about required water to live and reproduce. We do not know if Thales chose water as the basic stuff of the world for such reasons as these, but it certainly would be in character for him to have used such reasoning given his preference for calling on natural causal forces to explain things. Further, Thales' theory was maximally *unifying* in the sense that it sought to understand qualitatively different phenomena as being due to the same one underlying reality: water. That sort of unification of separate phenomena under a small number of basic explanatory concepts is one feature that distinguishes science from most forms of speculative tribal wisdom.

What Thales began other Milesians continued. Anaximander (a-naks-e-MAN-der) was another Milesian scientist and philosopher, only he claimed that all things stem from *to apeiron*, an untranslatable Greek phrase that usually gets rendered as 'the boundless'. The boundless might sound mystical and otherworldly to trendy modern ears, but Anaximander meant nothing especially supernatural by the term. It was a kind of neutral stuff, without magical powers, and it was certainly not a deity of any kind. Anaximenes (an-acks-SIM-e-knees), who came after Anaximander, holds an honored place in the history of science because of a vital improvement on Thales' and Anaximander's theories for which he was responsible. Thales had not provided a causal mechanism by which water gets transformed into things that appear not to be water. A piece of granite rock surely does not appear to be very waterlike. Hence, how can it be true that it is made of water? Thales apparently had nothing to say about this issue (or, if he did have something to say, we have lost the record of it). Anaximander had more to say on how the boundless is transformed into ordinary material objects, but his explanation did not involve a causal mechanism. The boundless is broken into parts in an act of "injustice," for which justice must be meted out by the eventual contraction of everything back into the boundless. Here one still sees a holdover from anthropomorphic modes of understanding the universe. Anaximenes, on the other hand, provided the first *naturalistic* causal mechanism by which a substance that possesses one bunch of qualities can be transformed into a substance with a different bunch of qualities. Anaximenes said that all things stem from air (not water or the boundless), and that all qualitative differences among things are due to different degrees of rarefaction and condensation of the air out of which all things are made; so, all qualitative differences are *reduced to* quantitative differences. Here was science to be sure. For the main job we expect science to do is to tell us, in nonmagical terms, what is really going on in the world—what there is in it, how what there is in the world got that way, and where what there is in the world will go next.

Twenty-five hundred years later we still expect science to tell us what is really going on in the universe. Some philosophers of science claim that how we measure progress in science comes down to the degree to which the several sciences are unified into one interconnected network of theories, a network that indicates that the universe is a cosmos and not a chaos, that it is in a nontrivial sense a single thing whose parts fit together into a coherent pattern. The term for the relationship between scientific theories that the positivists thought made possible such an interconnected network of theories is **reduction**. To reduce one theory to another theory was, under the positivist model of science, the ultimate goal of science. A reduction is naturally interpreted as achieving progress in an ontological sense. If, for example, chemistry is reducible to physics, as indeed it seems to be, then if we were to make an ontological inventory of all the kinds of objects and properties there are in the world, we would not in addition to all the physical kinds of objects and properties (that is, all the kinds of physical subatomic particles and their subatomic properties) need to list any chemical objects and properties. Why not? Because chemical objects and properties just are combinations of physical objects and properties. For example, the chemical property of *having a valence of +1* is "really" just the physical property of *being a single-electron donor*, that is, of how filled the outermost electron shell of the atom is. Chemical valence is not something distinctly chemical and nonphysical that must be added to our inventory. Instead of there being more basic objects and properties in the universe, there are fewer. And that is good, because the fewer the basic objects and properties, the lower our risk of error in doing science (there is less for us to be wrong about). The last sentence leads naturally to a principle of scientific method often called **Ockham's Razor** after its most famous advocate the medieval philosopher William of Ockham. Ockham's Razor holds that we should never multiply the number of independent entities and properties in our scientific theories beyond the minimum necessary to account for the phenomena in question, that we should never prefer a theoretical account of the universe that represents the universe as bloated with unnecessary entities and properties to a theoretical account of the universe in which it is a thin universe with just enough in it to appear as it appears to us. Ockham's principle can be used like a razor to pare or trim away from our scientific theories unnecessary posits and concepts, hence the principle's rather unusual name.

There have been a few successful reductions in the history of science. Thermodynamics in physics is reducible to statistical mechanics. In biology Mendelian genetics is reducible, or so some have claimed, to molecular genetics. The reduction of chemistry to physics has already been mentioned. There have also been a few what we might call partial reductions, reductions that have not come close to being completed but which have met with enough partial success to encourage further pursuit of a reductive research strategy. Some aspects of psychiatry seem reducible to neurophysiology, for example, so perhaps all of psychiatry is reducible to neurophysiology if we pursue that research goal with enough patience and stamina. Some proposed reductions have met with considerable controversy. The sociobiologists aim to reduce sociology (and perhaps even ethics!) to evolutionary biology. Needless to say, most sociologists are not especially fond of this proposal. In the 1970s some theorists sought to reduce cognitive psychology to neurophysiology plus the theory of artificial intelligence. This raised many hackles throughout the community of psychologists. It is interesting to note that this was one area of intense debate that had a direct impact on the philosophy of science. We will explore that impact in section 5.4. For now we need to ask the obvious question: precisely how is a reduction carried out? What is involved in reducing one theory to another? Different models of intertheoretic reduction have been offered to answer this question. We will look at two such models—the most influential ones in my view—in the next two sections.

5.1 Nagel's Model of Reduction

The philosopher of science Ernest Nagel occupied a prominent position among American philosophers for nearly thirty years. He trained several outstanding philosophers of science, of more than one generation, during his long tenure at Columbia University. Nagel was trained in the positivist model of science. He never completely abandoned it, but he nevertheless moved away from a strictly positivist approach in many respects. His monumental text *The Structure of Science*, published in 1961 (one year before Thomas Kuhn's antipositivist masterpiece appeared), was the culmination of the neopositivist attempt to salvage some essential core of positivist doctrine at the center of philosophy of science, a particularly heroic attempt in the face of the then newly arising historicist model of science that was eventually to be associated with Kuhn and his aftermath. Nagel's book is a bit imposing: thorough, technical, dry at times, but clearly written and written by someone who knows the philosophy of physics inside and out, the text presents many discussions that have since become classics of their respective subject matters. One of those classic discussions is the famous chapter on reduction, in which Nagel presented a model of reduction that was applauded for its clarity and precision even as it was skewered for its quixotic rigidity and severe formalism.

Nagel's model of reduction was tied to the positivist conception of scientific theories in the sense that reduction was applicable only to theories that were formalizable as first-order languages in the way the positivist conception demanded. A theory, in order to be reduced or to do the reducing, must, for Nagel, be expressible as a set of formalized sentences with the sort of logical structure we explored in chapters 2 and 3. The theory must have a set of core theoretical laws, and it must have a vocabulary distinguishable into two mutually exclusive classes of terms: theoretical and observational. Assuming that scientific theories can be so formalized, Nagel's model of reduction is remarkably clean in principle but extremely sloppy to pull off in practice.

Let T_1 and T_2 be formalized theories. Then T_2 reduces to T_1 if and only if the following conditions hold:

1. *The Condition of Connectability*: For every theoretical term 'M' that occurs in T_2 but not in T_1, there is a theoretical term 'N' that is constructible in T_1 but not in T_2 such that

for all objects x, x has M, if (and possibly only if) x has N.

The sentence isolated immediately above is called a **bridge law**, because it serves as a bridge to link two distinct linguistic expressions in two distinct theories. The requirement represented by the condition of connectability is that a bridge law must be provided for every theoretical term in the reduced theory that is not in the reducing theory. It is important not to misunderstand the condition of connectability. The condition does *not* require that 'N' *mean* the same as 'M' means. Nagel is careful to say that 'N' can either be coextensive with 'M' or else a sufficient condition for 'M'. Of these two possible relations, coextensitivity is the stronger, but it is not equivalent to sameness of meaning. To see the important difference between sameness of meaning and coextensivity, an example will be helpful. Researchers believe that the gene associated with the disease cystic fibrosis resides on chromosome 7. Suppose this is in fact the case, and suppose further that eventually the branch of specialized medicine that treats diseases like cystic fibrosis is reduced to some form of supergenetics. That would mean, if connectability was taken to involve coextensitivity, that there will be some bridge law to the effect that, for example,

for all organisms o, o has the cystic fibrosis gene if and only if o has DNA nucleotide sequence Z-2134 at location X on chromosome 7.

But clearly the expression 'cystic fibrosis gene' does not *mean* the same as the expression 'DNA nucleotide sequence Z-2134 at location X on chromosome 7'; for the latter expression mentions a chromosome and specifies a long sequence of nucleotide triplets and the former expression does not. The two expressions have different meanings, but they nevertheless are expressions predicable of all and only the same things, that is, of exactly the same things. Any organism to which one expression is applicable is an organism to which the other expression is applicable, and whatever organism one is not applicable to, neither is the other applicable to that organism. The technical term for expressions that are applicable to exactly the same things is 'coextensive'. To take another example, the two terms 'renate' and 'chordate' are coextensive. A renate is a creature that possesses a kidney, and a chordate is a creature that possesses a heart. It is simply a brute regularity of nature that every creature which has one has the other—there is no creature that possesses one but lacks the other. But, obviously, the two terms don't mean the same at all, for the meaning of each term involves a different type of bodily organ.

What Nagel's condition of connectability demands is that, for every theoretical term in the theory to be reduced that does not occur in the reducing theory, we would ideally want there to be some coextensive expression constructible using the vocabulary of the reducing theory, although short of that we will settle for an expression constructible in the vocabulary of the reducing theory that denotes a sufficient condition for the application of the reduced term. A good example would be the chemistry case mentioned above involving valence and electron donation. The predicate expression 'being a single-electron donor' is coextensive with the expression 'having a valence of +1' (not synonymous with it).

Now once we have all the bridge laws constructed we must meet one further condition to reduce T_2 to T_1:

2. *The Condition of Derivability*: It must be possible to derive logically all the theoretical laws of T_2 from the theoretical laws of T_1 together with the bridge laws. Let 'B' be the conjunction of all the bridge laws constructed in step 1, and let 'L_2' and 'L_1' be the conjunctions of the theoretical laws of T_2 and T_1, respectively. Then the Condition of Derivability requires that:

$$(L_1 \ \& \ B) \supset L_2.$$

The reader is reminded that the symbol ' $\supset$ ' represents the material conditional in first-order mathematical logic.

There is a refreshing clarity and precision to Nagel's model of intertheoretical reduction. The two steps have an intuitive appeal: First we must logically connect the terms of the respective theories by constructing bridge laws; then we must be able to derive logically the laws of the reduced theory from the laws of the reducing theory plus the bridge laws. What could be more elegant and to the point than that? It was a model that Nagel showed made sense of certain previously obtained reductions in physics, for example, the reduction of thermodynamics to statistical mechanics. But there were a couple of glitches in the otherwise rosy picture of reduction that Nagel had painted. One glitch is relatively minor and the other glitch is much more serious. I will discuss the minor glitch here and reserve discussion of the more serious one for section 5.4.

In the usual case, T_2 is not reducible to T_1 as T_2 stands, for T_2 is technically false as it stands. The reduced theory is often false in the sense that it only holds good in special circumstances. The reducing theory is the correct *general* account of the phenomena in question. For example, suppose it were possible to derive logically the laws of Newtonian physics from the laws of Einsteinian physics plus a suitable conjunction of bridge laws. That would be terrible news for physicists!

For Newton's laws are technically false, and so anything that logically implied them would be false also (in virtue of our old friend the logical rule of inference, *modus tollens*). Surely we cannot blame the bridge laws, given that they were specially and carefully constructed for the reduction in question. Therefore, we must take the reducing theory to be false. This is clearly an unwanted result. The solution to the quandary is to correct the errors in Newton's laws first; that is, we must correct the errors in Newtonian physics before we reduce it to Einsteinian physics. Physicists often say that Newtonian physics is just a "special case" of Einsteinian physics; Newtonian physics is what you get when you take the Einsteinian laws and apply them to systems whose velocities are well below the speed of light and whose masses are well below that of a stellar-size object. Well, yes and no. Yes, if you first "correct" Newtonian physics (for example, you must make mass dependent on velocity, and you must remove instantaneous causation at a distance). No, if you fail to correct Newtonian physics but instead take it as it is. But then we are caught in the following dilemma. If we correct the theory to be reduced beforehand, then we don't really reduce *it*, but only some corrected version of it; in our example, we don't really reduce Newtonian physics to Einsteinian physics, we reduce "Snewtonian" physics to Einsteinian physics (where 'Snewtonian' refers to what is really a different theory from the original Newtonian physics). On the other hand, if we fail to correct the theory to be reduced beforehand, then the reduction shows that the reducing theory is false (for it logically implies something false). So, either we failed to reduce the original theory we wished to reduce, or else the successful reduction in fact destroyed the truth of the reducing theory. Either we fail or else success is logically deadly. Neither result is a happy one for the reductively minded practitioner. In fact, the result is sufficiently unhappy that it spurred philosophers of science to construct a separate model of reduction.

5.2 Kemeny and Oppenheim's Model of Reduction

Two philosophers of science, John Kemeny and Paul Oppenheim, argued that the Condition of Connectability in Nagel's model of reduction is almost never fulfilled in what are otherwise considered successful reductions. The reductionist is almost never able to construct term-by-term bridge laws, for the holistic interdependence of theoretical terms defeats the attempt to do so. The bridge laws become so huge and unwieldy as to be useless, if not meaningless. So, Kemeny and Oppenheim built a model of intertheoretic reduction that contained neither of Nagel's conditions; for you can't meet the Condition of Derivability without having first successfully met the Condition of Connectability (you can't logically derive the laws of the reduced theory minus bridge laws that connect the two theories in question). Kemeny and Oppenheim argued that what matters in a reduction is that the reducing theory be able to *explain* any observational data that the reduced theory explains, that the reducing theory be in some genuine sense different from the reduced theory, and that it be at least as "systematic" as the reduced theory. Their model also did not try to provide necessary conditions for a successful reduction, but only sufficient conditions. Again, let T_1 and T_2 be formalized theories. Then, under the Kemeny and Oppenheim model of reduction, T_2 is reducible to T_1 if

(1) T_2 contains terms not in T_1
(2) any observational data explainable by T_2 are explainable by T_1
(3) T_1 is at least as well systematized as T_2.

Condition (1) ensures that there is an ontological point to reduction. Condition (2) makes explanatory power a key factor in the evaluation of intertheoretic reductions. Condition (3) is required, said Kemeny and Oppenheim, to rule out "wrong-way" reductions. For example, reducing psychology to astrology is forbidden because astrology is theoretically more disorganized, more unsystematic, than psychology.

The fatal flaw in the model is the third condition. It would certainly appear to be the case that some reductions involve reducing a simpler—that is, a better systematized—theory to one that is more complicated and less well systematized precisely because the reducing theory doesn't oversimplify the target phenomena where the reduced theory does oversimplify the target phenomena. Consider the reduction of chemical theory to quantum physical theory. This is surely a reduction going in the correct direction; yet, it is just as surely a reduction of a fairly well systematized body of theory to one that is notoriously, if not infamously, conceptually problematic and simply not as well systematized. Are fundamental subatomic entities particles or waves? They are both, says Niels Bohr? But a wave is an entity spread across more than one point of spacetime and a particle is an entity not spread across more than one point of spacetime, we object. How is this moving from a fairly well systematized macroscopic chemistry to a theory of quantum phenomena at least as well systematized? The answer is that, of course, at present it is not any such movement at all; for the history of quantum physics up to now suggests that the ultimate microstructure of nature is inherently indeterministic and sloppy (there are currently over 200 distinct subatomic wavicles), and on those grounds alone quantum physics may be doomed to experience a kind of permanent upper limit on how well systematized it can become—a limit surely exceeded by many of the higher-level theories we seek to reduce to it. Of course, this has not discouraged quantum physicists from seeking to unify quantum phenomena through reduction to some small set of fundamental objects and properties, but the point is that methodologically they have no grounds for being guaranteed up-front that their attempts at unification must succeed.

To take an example closer to our chosen domain of inquiry in this text, consider the recent attempts to begin reducing what were previously thought to be unrelated diseases to a theory that treats them all as autoimmune diseases. My prediction is that in 50 to 100 years autoimmune medicine will be a separate specialty. It will achieve such an independent status by "stealing" for its focus of inquiry phenomena from a number of currently distinct fields of medicine. The following is a list of some of the diseases known to be autoimmune in nature and the medical fields that are currently responsible for studying and treating them.

1. systemic lupus (rheumatology)
2. insulin-dependent diabetes (endocrinology)
3. myasthenia gravis (neurology)
4. multiple sclerosis (neurology)
5. Grave's disease (endocrinology)
6. rheumatoid arthritis (rheumatology)
7. bullous pemphigoid (dermatology)
8. scleroderma (dermatology)
9. ulcerative colitis (gastroenterology)
10. pernicious anemia (hematology)
11. Addison's disease (endocrinology)
12. Goodpasture's syndrome (nephrology)
13. primary biliary cirrhosis (gastroenterology)

The list could be expanded to greater length, but my argumentative point should be clear from the listed examples: Given that the theory of autoimmune phenomena must account for physiological disease processes in so many different kinds of organs and tissues, under so many different kinds of local conditions, the theory of autoimmune diseases is bound to be more complicated, less organized (at least at first), and generally less well systematized, than the various rather narrowly focused specialties noted in parentheses. The point I'm making against the Kemeny and Oppenheim model of reduction can be put another way. It isn't *intra*theoretic systematization that matters to us, or to the reductively minded practitioner, it is *inter*theoretic *unification* that matters. We don't care so much whether a theory is internally sloppy and disorganized as much as we care whether that theory has the potential to unify—to reduce to the same collection of terms and concepts—phenomena from a number of domains of inquiry and practice previously thought to be distinct and unrelated to each other. Practitioners will pursue a reductive research strategy—even one with a horribly unsystematized theory as the reductive base—if they have grounds for believing that the unsystematized base theory is nevertheless the one dealing with the real causal and ontological aspects of the phenomena at issue. For example, consider the continuing research into neurophysiology as a reductive basis for all psychiatric phenomena despite the notoriously unsystematized, dare I say entirely conjectural, state of much neurophysiological theorizing at the present time. The gargantuan problems associated with simply localizing psychological functions in the nervous system may set the same sort of upper limit on being well systematized that potentially plagues quantum physics.

In summary, what really motivate the reductively minded practitioner are causal and ontological issues, not issues of explanation and systematization. Kemeny and Oppenheim were still operating within a positivist mind-set, one that sought to eliminate ontological motivations from science and replace them with epistemological substitutes. Instead of having reductively minded practitioners care about what causes what out in the real world, Kemeny and Oppenheim have them care mainly about how tidy and spiffy their theories are. But the large majority of scientists do care, and care very much, about what is going on in the universe—or at least in the corner of it that they investigate. They are the philosophical and temperamental descendants of Thales, Anaximander, Anaximenes, and other Milesians of 2,500 years ago. Like Thales, Anaximander, and Anaximenes, most scientists are ontologists at heart. They wish to understand the cosmos, and to understand it *as* a cosmos, and that means that they wish to understand how its parts are unified into a cosmos. This presumption that the universe is a unified system of domains that form one hierarchical network is called the **Unity of Science**. A closer look at the Unity of Science research program will help us to understand why reductionism, despite its recent decrease in popularity, is a permanent feature of the landscape within philosophy of science.

5.3 The Unity of Science Program

The Unity of Science was a concept developed within the positivist tradition in philosophy of science. The esteemed American philosopher Hilary Putnam and, once again, Paul Oppenheim, coauthored an influential study of the Unity of Science program in 1958 (although now Putnam would surely disown the notion). The idea at the core of the Unity of Science is deceptively simple: *There are no special sciences*, where by 'special' one means a science whose fundamental concepts are somehow incapable of being connected with or related to the fundamental concepts of at least one other science. For if there was such a special science, then presumably the phenomena it investigated would also be phenomena somehow determinatively unconnectible to any other kind of phenomena in the universe. The universe in such a case would contain an unexplainable discontinuity in it. It would fail to form one continuous fabric "all the way down" from the largest supergalactic superstring to the smallest subatomic particle. The universe in that sense would fail to be a cosmos. But how could that be? How could it be that certain theories from different domains "line up" in reductive order, but that this is merely a local achievement—that there are theories that must remain outside the reductive network no matter how far science advances? Such a result is metaphysically obnoxious, if not downright bizarre. The bizarreness of special sciences can be brought out clearly if we read the lessons of current astrophysical cosmology correctly.

Cosmology is the branch of astrophysics responsible for giving an account of the current overall structure of the universe and how it got that way up to the present. Such a task requires tracing the evolution of the universe from its beginning, and that tracing involves understanding the best available theory of the origin of the universe—the misleadingly named big bang theory. If the universe began from a singularity that "exploded," as the big bang claims it did, then the universe is bound to be a completely homogeneous entity; that is, there would be no ontological grounds for some parts of the universe being so qualitatively distinct that they can't be fit in theoretically with any other parts of the universe. To use a crude analogy, if a cake starts out as flour, eggs, butter, milk, chocolate, and salt, then how could what comes out of the oven as the finished product be something with certain properties in one part of it that are wholly incompatible—logically, causally, and theoretically incompatible—with the properties in any other part of it? After all, its all just so much flour, eggs, butter, milk, chocolate, and salt, which have been mixed together and heated.

To return to the main path of our argument, a special science would be a science of special phenomena, phenomena that destroyed the homogeneity of the universe in virtue of their specialness. The positivist philosopher Herbert Feigl once called the theoretical laws about such special phenomena, the laws that would presumably be at the core of any special science that investigated such phenomena, **nomological danglers**. 'Nomos' is the Greek word for 'law'. Such laws would "dangle" outside the rest of the scientific network of theories, and therefore they are to be rejected for three reasons. First, they violate Ockham's Razor. Recall that Ockham's Razor requires that we do not multiply distinct entities and properties beyond the minimum necessary to explain the phenomena in question. Second, nomological danglers are inconsistent with the continuity of nature, a fundamental principle most scientists embrace. Why do they embrace it? Because of the third reason to reject nomological danglers: They are conceptually incompatible with the big bang cosmology, a cosmology that entails the continuity of nature.

Putnam and Oppenheim in their influential paper presented an argument for the Unity of Science that involved mereological premises. Mereology is the study of part/whole relationships. The two premises in question they called the *principle of evolution* and the *principle of ontogenesis*. The first principle holds that the universe evolved from smaller to larger in terms of organization; so, for any level of organization in nature, there was a time in the evolution of the universe when there were systems of that level but no systems of any higher levels of organization. The second principle holds that for any particular system at a given level of organization, there was a time in the evolution of the universe when the system did not exist but some of its parts on the next lowest level of organization did. Both these mereological premises suffer from serious problems. The principle of evolution assumes that there is a natural and unique way to decompose the universe into "levels" of decreasing organization, that there is one correct way to do this. However, there are different ways of cutting up a system into parts. It all depends on what your purposes are in decomposing the system. The principle of ontogenesis assumes something even more questionable: that it makes sense to speak of a system's parts existing before the system itself exists. But there are notorious identity problems with this way of proceeding. It would seem to be a safe bet that a system must either exist or have existed in the past for us to speak of some other thing being a part of that system. In other words, the prior existence of the system is a necessary condition for the existence of any of the system's parts. The property of being-something-which-will-be-a-part-of system S is not the same as the property of being-a-part-of system S. If this is so, then the principle of ontogenesis suffers from a fatal internal incoherence.

I do not think that the proponents of the Unity of Science program needed to base it on any mereological principles like those Putnam and Oppenheim offered in 1958. The mereological principles have a contrived air about them, a whiff of neopositivist desperation. The real motivations for belief in the Unity of Science research program have always been twofold: past reductive successes, and the continuity of nature as an implication of the big bang cosmology. These were the real epistemological grounds for the Unity of Science, not a series of arcane principles about parts and wholes. It will help to have a pictorial representation of the Unity of Science to understand the main critique of it that many philosophers of science believe killed it off for good. Even if those philosophers of science are wrong, and the Unity of Science is not dead, it still had to be modified enormously to defend it adequately against the critique. That modification made the Unity of Science nearly unrecognizable with respect to its original form. The modification changed it to a "nonreductive" Unity of Science, and we shall investigate this nonreductive form of the Unity of Science in section 5.5.

Figure 5.1 hopefully presents the Unity of Science program in a way helpful to the reader. To facilitate the coming critique of the Unity of Science I have split the higher-level sciences into the two classes depicted. The idea is that both physiology and geology reduce to different subbranches of chemistry. Note the converse directions of determination (for example, causation) and explanation. Determination moves from lower to higher and explanation moves from higher to lower. Crudely, parts determine wholes, and wholes are explained by their parts. Whether the ancient Pythagorean philosophers were correct and the universe is ultimately a purely mathematical entity is a subject of current dispute, so I have put a question mark next to 'mathematics' in the core sciences box. Certainly one can understand why the Unity of Science program as set out in Figure 5.1 might be attractive to a philosopher who is sympathetic to science. The notion of a universe that is a cosmos is encapsulated in this picture. The various strands of science all fit into one fabric, and their specific systems of concepts boil down to one architectonic scheme of categories. There are no special

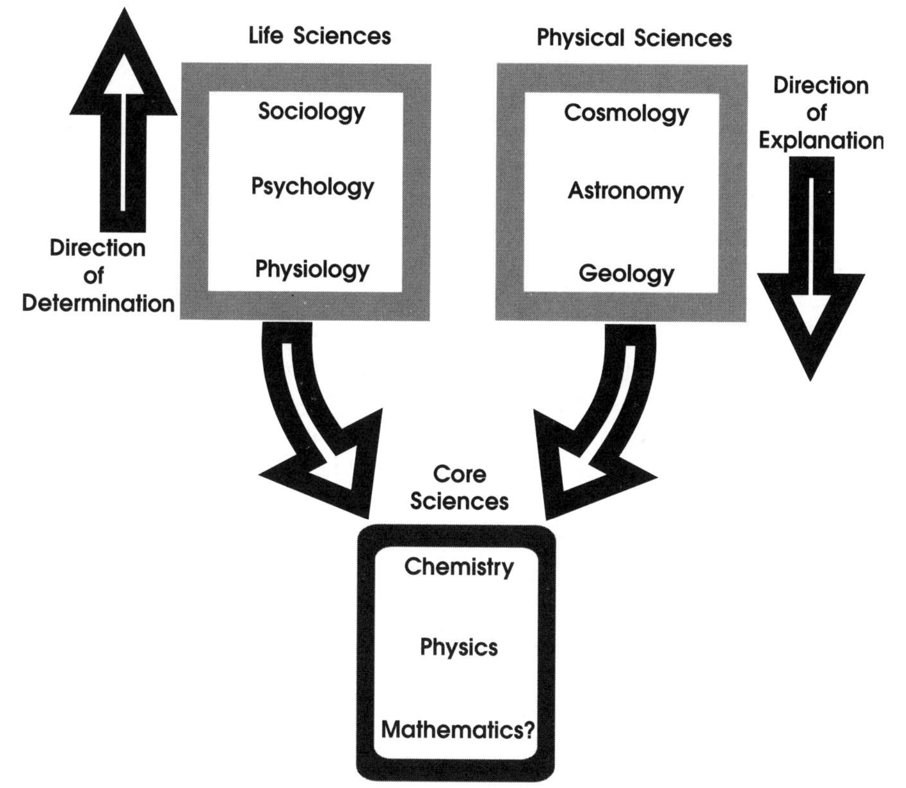

Figure 5.1. A version of the Unity of Science.

sciences. Unfortunately, as we are about to see, the picture is an oversimple distortion of what is really a much sloppier and murkier reality.

5.4 The Putnam/Fodor Multiple Realization Argument against the Unity of Science

A sly reader will have noted that our discussion of the Unity of Science research program remained at a disturbingly nondetailed level of description. It was all a bit too programmatic, with too many broad strokes and not enough narrow ones. How precisely would psychology connect up reductively with neurophysiology? How exactly does oncology (the medical science that studies cancer) reduce to immunology? Well, says the partisan Unity of Science supporter, those are mere details to be left to the grunt practitioners down in the trenches of research to figure out—a likely story, replies the Unity of Science skeptic, if ever there was one.

Putnam himself came to reject the Unity of Science Program only a few years after he'd written with Oppenheim the classic presentation and defense of it. Putnam made his rejection of the Unity of Science turn on the alleged nonreducibility of concepts within theories. The philosopher of psychology Jerry Fodor extended Putnam's point and turned it into a critique of the Unity of Science in which the incompatibility that destroys the possibility of reduction is an incompatibility between the actual phenomena the theories are about. The position that Fodor and Putnam both argued for in this way is called antireductionism. If Fodor is correct, then psychology is a special science in the strongest sense possible. What Fodor argued was that there exists a particular level of organization in nature at which a discontinuity occurs between phenomena of different types; so there is a theoretical discontinuity between the sciences that supposedly meet in the reductive hierarchy at that level of organization. Where is this disruption in the fabric of nature? It is where psychological phenomena meet neurophysiological phenomena. So, the theoretical discontinuity, the place where reduction hits a dead end long before it should, is where psychology meets neurophysiology (where psychology meets physiology in Figure 5.1).

To oversimplify Fodor's charge to the point of near vulgarity, the mental ain't merely the physical. To be sure, Fodor hedges and qualifies his claim (using what strikes some philosophers of science as a truly hair-splitting distinction without any real force), but in a way that seems to strip it of any interesting power and which it would not profit us to consider in greater detail here.

Fodor's argument depends on the viability of a very controversial philosophical concept: **natural kinds**. The concept of a natural kind comes from the ancient Greek philosophers, especially from the works of Aristotle, the first systematic biologist in the history of science. Aristotle claimed that science seeks to uncover not merely the accidental aspects of the phenomena it studies, but its *essential* aspects as well. For an Aristotelian biologist, the theory that investigates carnivores, for example, must uncover the **essence** of being a carnivore. A set of entities capable of having an essence forms a *kind*; and, if the entities in question are naturally occurring entities, the kind is a natural kind. All carnivores, to use our example, must share some essential property or properties in virtue of which they are all members of the natural kind denoted by our term 'carnivore'. What might such a kind-forming essence be like? Well, how about this: *Exclusively meat-eating* is the essence of a carnivore. Hence, carnivores are a natural kind, and they therefore figure in the causal network of the biological world. For example, carnivores causally interact with noncarnivores of various natural kinds: herbivores and omnivores, for instance. Furthermore, carnivores themselves come in various natural kinds: lion, hyena, jackal, cheetah, and so on. Each of these natural kinds has its own kind-determining essence. In physics there are natural kinds: electron, proton, neutron, lepton, meson, quark, and many others. Within immunology there are natural kinds. NK cells form a natural kind, one whose essence might be given as, *non-MHC–restricted, nonsensitized killer of altered-cells* (the reader need not worry for now about the technical modifying expressions—take my word for it, the essence of an NK cell, if it has one, is something close to this).

What natural kinds allow a practitioner to do, among other things, is to cut the domain of inquiry along its natural "seams." For example, no biologists practice their science as though there exist objects of investigation within their chosen domain of inquiry that are called "zebons," where a zebon is a mammal composed of the front half of a zebra and the rear half of a lion. There simply is no such kind of critter in nature. Notice certain epistemological and methodological implications that flow from the absence of a natural kind: There are no theoretical laws about zebons, and so there can be no predictions about them. Likewise, zebons do not participate in any causal relationships with other natural kinds that actually exist; so, zebons cannot figure in any explanatory relations involving natural kinds that actually exist.

That the universe presents itself to us with natural kinds already in it is a central presumption many scientists make, perhaps most of them unconsciously so. That a theory's fundamental laws should be about the natural kinds that occur in that theory's domain is another central ideal at which many scientists aim, again perhaps without an explicit awareness on the part of many of them that they do so. Jerry Fodor certainly presumed that psychologists are among such scientists. Like any other domain of inquiry, psychology seeks to uncover the genuine causal relations between the genuine natural kinds that form the seams of the psychological level of phenomena within the universe. Psychological laws of nature will relate natural kinds to each other, natural-kind properties to each other, and natural-kind events to each other. But that means, argued Fodor, that psychology cannot be reduced to any theory or collection of theories from any sciences below it in the hierarchy presumed by the Unity of Science; for when we understand how it is that psychological states and properties of organisms are physiologically realized in the nervous systems of those organisms, we shall see that *the terms constructed in the reducing theory to form the bridge laws, in many cases, will not denote natural kinds.*

Why not? Because a given psychological state or property is "multiply realizable" in the physiology of a typical nervous system. Take the psychological state of *believing that there is life after death.* There will be no *one*, single, physiological way for that state to be realized in the neurophysiology of a human being. Suppose Casper and Daphne both believe that there is life after death. There may be nothing physiologically similar about the way that mental state is represented in the nervous systems of Casper and Daphne. Yet, in each of their nervous systems, there is *some* neurological representation of a mental state with that specific content: believing that there is life after death. Suppose we begin to survey the nervous systems of people who each believe that there is life after death. We may find that out of, say, 40,000 such people, there are a total of 28,374 *different* ways that particular mental state can be neurophysiologically realized in the human nervous system. But this destroys the possibility of constructing a bridge law linking the mental predicate 'believes that there is life after death' with a coextensive predicate in neurophysiological theory, claims Fodor; for the half of the bridge law representing the neurophysiological term would be an artificial disjunction—a huge predicate composed of separate neurophysiological descriptions disjoined to each other (linked by 'or')—that *does not denote a natural kind.* But laws of science link natural kinds, not artificial kinds, so the bridge law would not be a law at all, but rather some monstrously complicated and predictively useless correlation between a natural kind mental state and . . . what should we call it, a term that refers to no piece of nature?

Let 'M' denote the mental property of believing that there is life after death. Let 'NP_1', 'NP_2', 'NP_3', and so on denote each of the 28,374 different neurophysiological realizations of the psychological property of believing that there is life after death. Then the bridge "law" needed to link this mental state with the neurophysiological realm would look like this:

for all organisms o, o has M if (and possibly only if)

(o has NP_1 or o has NP_2 or o has NP_3 or o has NP_4 or o has NP_5 or o has NP_6 or has . . . or o has $NP_{28,374}$)

The last half of this statement does not pick out a natural piece of the universe. Physiological state NP_{64} may be nothing like physiological state NP_{567}, which in its turn may be nothing like physiological state $NP_{7,441}$, and so on. The individual states connected by the or's have no one thing in common. Hence, the entire last half of this supposed bridge law is a specially concocted disjunctive term that only exists because some crazy reductionist wants to reduce the psychological level of nature to the neurophysiological level of nature. The statement is predictively useless; for we would not know which of the 28,374 disjuncts was true of any given person who believes that there is life after death. All we could say of any given case is that some disjunct or other must hold good. But that is so predictively useless as to be a meaningless exercise. But if the statement is predictively useless, it is also explanatorily useless.

The reader may wonder why there are so many different ways in which the same mental state could be realized within the nervous system of a human being. There are good grounds for thinking that this is so, but they involve some rather technical distinctions and arguments within the philosophy of mind, distinctions and arguments not directly relevant to our purposes in this text. Perhaps it is sufficient here simply to note that the content of many mental states is characterized *functionally*; that is, their content is characterized causally. Mental state M is the mental state of believing that p, or wishing that p, or whatever, because it is that state which is caused by certain input stimuli and that in turn causes the organism to produce certain output behavior. If this is correct, then the content of a mental state has nothing to do with the particular kind of physical material in which it is represented. If being angry at the current U.S.

president is the mental state with *that* content—it is anger (not joy or indifference) directed at the current U.S. president (not your brother, not a past U.S. president)—in virtue of being that state of the organism caused by input experiences e_1 through e_n (for some number *n* greater than 1) and which in turn causes the organism to produce output behavior b_1 through b_n, then *any* organism for which e_1 through e_n is connected with b_1 through b_n in this way is an organism angry at the current U.S. president, *no matter what that organism is made of, no matter what its internal structure is.* If space aliens land in a spaceship tomorrow and it turns out that they too are angry at the current U.S. president, that is perfectly possible even if the space aliens turn out to have a physiology of boron, nickel, tin, and silicon (instead of a human hydrocarbon physiology).

Suppose you and I are both thinking of the first time human beings landed on the moon. This is a mental state functionally characterized. That means that there doesn't have to be anything, not one single thing, similar in your brain and my brain in virtue of which we are both in this same mental state. That is why Fodor's argument against reductionism gets off the ground: *The multiple physical realizability of mental states will result in bridge laws that aren't laws at all because their physiological sides do not pick out natural kind properties, objects, or events in the physiological domain of nature.*

The Putnam/Fodor multiple realization argument created much consternation in philosophical circles. Some reductionists tried to duck its impact with a "so what?" response. Other philosophers jeered at the argument's use of the concept of natural kinds on the grounds that there are no such things; or worse, on the grounds that there are natural kinds but we get to say what they are, not nature (and so, the last half of such a bridge law is as natural as you want it to be). Still other philosophers took the argument to heart and said good-bye to reductionism. There is at least one special science after all—psychology.

To most philosophers none of the above reactions seemed adequate. The argument must be given its due respect, but neither must the baby be thrown out with the bathwater. One such thinker was the philosopher Jaegwon Kim. Kim sought to recognize the multiple realizability of mental states while at the same time he wanted to rescue the ontological point of reduction. The main philosophical purpose of reduction is to achieve an ontological economy of sorts, to show that there are not two ontologically distinct kinds of phenomena going on, but only one kind of phenomena going on. The mental just is the neurophysiological, at least it is for human beings. Kim achieved his reconciliation by revamping a heretofore ignored kind of asymmetric covariance among distinct families of properties, an asymmetric covariance that is due to the fact that one family of properties is determinatively dependent on a qualitatively distinct family of properties without properties of the former family necessarily being reducible to properties of the latter family. Further, the determinative dependence at issue is not necessarily ordinary causation; for ordinary causation is lawlike and hence holds only between natural kinds (and therefore would support a reductive relationship between cause and effect). This interesting brand of asymmetric dependence Kim called **supervenience**, and to its consideration we now turn.

5.5 Supervenience to the Rescue?

When the broad-spectrum antibiotic penicillin was first approved for use by physicians it was prescribed as though it was a wonder drug. Doctors used it to treat every minor sore throat and puffy big toe. Medical microbiology was uncertain at the time precisely how an antibiotic kills its target microorganisms, and that ignorance of detail was filled in with an unsupported impression that penicillin killed practically anything small enough to require a microscope to identify. The resulting overprescription produced, among other results, a host of penicillin-resistant strains of microorganisms that otherwise would not have evolved (or, at least, they would not have evolved as soon as they did). It is perhaps in the nature of scientists to seize on a new instrument, material, or concept in the hope that it is a universal tool capable of solving the most severe theoretical, experimental, and practical problems that hound daily practice in their fields.

Philosophers are only slightly less trendy in the practice of their own chosen field. New concepts and arguments are often seized on by philosophers in the hope that they are tools that at last will settle some long-standing dispute or render clear some cloudy issue of perennial endurance. Supervenience has suffered just such a fate. It became the "fab" concept in metaphysics and philosophy of science in the years after Kim introduced a precise model of it. But just like in the case of penicillin, its popularity rode a wave of vagueness, imprecision, and sometimes outright distortion. Many philosophers helped themselves to supervenience without first taking the time either to construct a precise model of it or to learn such a model invented by someone else. What supervenience is to one philosopher is often not the same as what it is to the philosopher up the road a piece. As a result there has been a proliferation of different concepts all claiming to be concepts of supervenience. This situation is not unlike a case in which fifty different psychiatrists present fifty different conceptions of what psychoneurosis is, or what schizophrenia is. Kim himself has noted the lack of a prior ordinary-language use for the terms 'supervene', 'supervenient', and 'supervenience', and that in such a situation these technical terms mean pretty much what the philosophical specialist wants them to mean. In the absence of corrective pressures emanating from ordinary speech, the main constraint on our use of a specialized term of art is its degree of pragmatic usefulness "under fire." What are we to do in the face of this profusion of models of supervenience? The present author was a student of Kim's, and he believes that Kim's original model of supervenience is the most useful and conceptually richest model of the lot of them. Accordingly, in this text we will understand supervenience as it is characterized in Kim's model of it, to which we now turn.

5.5.1 Kim's Model of Supervenience

The original purpose behind supervenience was a metaphysically deliberate one: to construct a concept of asymmetric dependence under which the determinative relation represented by the concept in question is: (1) not necessarily causal (it does not require lawlike regularities); (2) nonreductive (to allow for the many-to-one aspect of multiple realization); and (3) ontologically significant (strong enough to support, say, a general scientific physicalism, without being so strong as to be reductive). In short, what supervenience was originally designed to save from the aftereffects of the multiple realization argument was *noncausal, nonreductive physicalism.* 'Physicalism' denotes what used to be called materialism, the view that the universe is ultimately an entirely physical system. What physicalism implies is that a mild version of the Unity of Science holds good: Ultimately there are no phenomena in the universe which cannot be understood in terms of the concepts of physics. So, in a way, supervenience was called on to salvage a toned down version of the Unity of Science, a version in which the respective sciences line up in order of determinative dependency without lining up in reductive order.

The first tricky aspect of supervenience is that it is a relation that holds between natural groupings or what Kim called *families* of properties. One way of understanding this is to note that competent supervenience theorists rarely talk about a particular property supervening on another *particular* property (except as a limiting case for illustrative purposes only). They always talk in more general terms such as, "the mental supervenes on the neurophysiological" or "allergic phenomena supervene on immunological phenomena." The former statement can be interpreted along the lines in which the latter is expressed: mental properties and events (as a family of such) supervene on neurophysiological properties and events (as a family of such). Kim constructed his model of supervenience originally for families of monadic properties (that is, properties that are not many-place relations). Let the family of psychological properties attributable to higher organisms be represented by M*. The point of the '*' after the 'M' is to indicate that we include in M* properties formable by applying simple logical operations to the basic properties of M*, operations such as conjunction, disjunction, and negation. Let the family of neurophysiological properties attributable to higher organisms be represented by N* (where '*' again indicates simple compounding operations to make more complicated properties out of simpler ones). Then the psychological supervenes on the neurophysiological if and only if the following holds:

> necessarily, for any two objects x and y, if x and y have exactly the same N* properties, then they have exactly the same M* properties; and, if x and y differ in M* properties, then they differ in N* properties.

Fans of metaphysics should note that supervenience comes in different strengths—three, to be exact: global, weak, and strong. The version above is of middle strength, called weak supervenience. Arguments can be heard that weak supervenience is not robust enough to qualify as an interesting resolution of either the traditional mind-body problem, or the Putnam/Fodor multiple realization argument. I would prefer to use strong supervenience, but its use here would require the introduction of possible-world semantics into the discussion, a highly technical and complicated detour we cannot afford the chapter space to pay for.

Supervenience as applied to the case of psychology comes to this: There is no difference in psychological properties without a difference in neurophysiological properties, but there can be a difference in neurophysiological properties without there being a difference in psychological properties. This last remark should be very significant to the reader. It indicates that the psychological can supervene on the neurophysiological in a one-to-many fashion: The same psychological state can be neurophysiologically realized in many distinct ways. What cannot happen if the psychological supervenes on the neurophysiological is that the same neurophysiological state can be realized as more than one distinct psychological state. The neurophysiological varies if the psychological varies, but the psychological does not necessarily vary if the neurophysiological varies. The determinative dependence is *asymmetric*. The core question then becomes: Is asymmetric covariance a relation strong enough by itself to sustain a watered-down Unity of Science program?

5.5.2 Modest Physicalism

Kim has recently suggested a negative answer to the question that closed the last section. The results of work by many supervenience theorists in recent years indicate that asymmetric covariance is logically independent of asymmetric determinative dependence: Two families of properties can stand in a relation of supervenient asymmetric covariance, and it can be the case that neither is determinatively dependent on the other. But determinative dependence would surely be one minimum requirement of any Unity of Science program. If the various sciences cannot be lined up in reductive order, then at least they should line up in order of determinative dependence—they should, that is, if the Unity of Science is to have even the slightest interest. If Kim is correct, then it is a separate task to show that a form of determinative dependence underlies a given supervenience relation. Simply pointing out that supervenience holds will not be enough.

Suppose that allergic properties supervene on immunological properties. This is a supposition with much evidence behind it. Suppose further that immunological properties supervene on genetic properties, for which again, there is much evidence in support. Suppose further that genetic properties supervene on physicochemical properties, a supposition that has by now become part of standard scientific knowledge of the biological world. Once we have hit the physicochemical domain, the rest of the determinative order is pretty much set all the way to fundamental particle physics. Determination at that point is pretty much mereological in nature. Kim has argued that any supervenience relation worth the name is *transitive*. If A supervenes on B and B supervenes on C, then A supervenes on C. The presence of an intervening third family of properties in the supervenience chain from A to C may serve to illuminate the determinative ground of the supervenience relation of A on C. If this is so, then we at least have rendered plausible a "modest" version of physicalistic Unity of Science. But we need to note one important caveat. Supervenience, in Kim's view, is not correctly thought of as a self-contained "kind" of determination at all. It is rather a pattern of asymmetric covariance, and we may legitimately ask of any such relation of asymmetric covariance: What is the *determinative ground* of the supervenience relation? The idea is that in different domains of inquiry, a different kind of determination may ground, account for, the supervenience relations that hold in that domain. In chemistry and physics, for example, the supervenience of the chemical on the physical is determinatively grounded by mereological relations: The properties of larger wholes are determined by the properties of their constituent parts. The chemical properties of molecules, for example, are determined by the physical properties of their atomic constituents. Now in cognitive psychology and neurophysiology, it is presumably an open question subject to controversy just what kind of determinative relation grounds the supervenience of the psychological on the neurophysiological. Since the advent of the multiple realization argument, most theorists working in this area grant that the determinative relation cannot be causal (for there can't be the kinds of laws liking natural kinds required for causation). But what the determinative relation is in a positive sense is under dispute. The behaviorist, for example, might argue that the determinative relation in question is semantic: Psychological states of organisms supervene on physically characterized behavioral states of such organisms because of certain "analytic" semantic connections that hold between the linguistic expressions that denote psychological states and the linguistic expressions that denote physically characterized behavioral states. The supervenience is determinatively grounded in relations of semantic meaning. The functionalist psychologist would, of course, take a slightly different line of approach.

What the demand that we distinguish supervenience from its ground amounts to is an admission that supervenience does not itself constitute an explanatory basis for anything. The existence of a supervenience relation between property family A and property family B suggests that there is a determinative priority between the two families of properties; but it does not indicate the specific nature of that determinative relationship (is it mereological, causal, semantic, or some other kind of determination?). And because it does not indicate the specific nature of the determinative relationship, supervenience does not by itself count as an explanatory relation between the two families of properties. Supervenience suggests that there is some determinative relation between the property families in question that is sufficient to underwrite an explanatory relation between them; but the kind of determination may be different in different domains. Does it follow from this that the kinds of explanatory relations that obtain between different phenomena will also vary from domain to domain? Is what counts as a scientific explanation radically different in different domains? Or is there a single unified model of what is involved in explaining scientifically a particular phenomenon? The traditional view, the one that flowed naturally from the positivist model of science, gave an affirmative answer to the last question and a negative one to the others. Nowadays, however, there is less acceptance on the part of philosophers of science of such a monolithic conception of scientific explanation. We begin our own exploration of the thorny topic of scientific explanation in the next chapter.

Further Readings

Ernest Nagel presents the mature version of his model of reduction in chapter 11 of *The Structure of Science*, New York, Harcourt, 1961. J. Kemeny and P. Oppenheim discuss the technical details of both Nagel's model and their own brand of reduction in "On Reduction," *Philosophical Studies* 7 (1956): 6–19. P. Oppenheim and H. Putnam introduced a self-consciously ontological model of the Unity of Science in their classic paper "The Unity of Science as a Working Hypothesis," in *Minnesota Studies in the Philosophy of Science Volume II*, edited by H. Feigl, M. Scriven, and G. Maxwell, Minneapolis, University of Minnesota Press, 1958, pp. 3–36. Jerry Fodor unveiled the multiple-realization argument against psychophysical reductionism in the introduction to his *The Language of Thought*, Cambridge, Mass., Harvard University Press, 1975. An earlier version of this brand of antireductionism can be found in Hilary Putnam, "The Nature of Mental States" in his *Mind, Language and Reality: Philosophical Papers Volume 2*, Cambridge, Cambridge University Press, 1975, pp. 429–40. Nearly all Jaegwon Kim's important papers on supervenience are now collected in his *Supervenience and Mind*, Cambridge, Cambridge University Press, 1993.

6

The Structure of Scientific Explanation

The first philosopher of science to write with systematic deliberateness on the topic of scientific explanation was Aristotle (about 320 B.C.E.). It was Aristotle's view that explanation in science unveils the essence or being (the *ousia* in Attic Greek) of the phenomenon explained. Put in a slightly different way, he held that explanation in science requires correct "definition" of the phenomenon explained. This was unfortunate for two reasons. First, it entangled science too extensively with speculative metaphysics. The scientist was required not merely to give an adequate accounting of how we happen to experience the phenomenon being explained—to "save the phenomenon," as the saying goes—but rather the scientist was required to expose the "essential being" of the phenomenon. Second, the Aristotelian model of scientific explanation mixed explanation up with the arbitrary conventions of language insofar as it treated explanation as a matter of how we classify things using a collection of conventional categories, or worse, of how we arbitrarily choose to define words. But what is explained is something in the real world of experience—an event, an object, or a property that occurs, exists, or is exemplified—not a word or expression in a language. Yet only linguistic entities like words or expressions can have definitions. In chapter 4 we were introduced to an interesting type of immune cell called a natural killer (NK) cell whose main function is to identify and destroy cells of the host organism that have become transformed through cancer or viral invasion.

Now I do not explain to you *why* nor *how* NK cells kill cancer cells by defining the expression 'NK cell'. On the other hand, if I do explain to you why and how NK cells kill cancer cells, I am not required in the course of doing so to provide you with the metaphysical essence of an NK cell; for their capacity to kill cancer cells might be due to accidental, nonnecessary properties NK cells possess.

The view that scientific explanation involved appeals to essences and definitions was subsequently a great hindrance to the progress of science. One unfortunate feature of ancient science that historians have puzzled over was the negligible role it gave to experimentation. There was staggeringly little experimentation done in ancient science, and what little experimentation that occurred was unsystematic and haphazard: There was no general theory of experimental procedure. I believe that we should not be so puzzled about the lack of experimentation in ancient science once we understand how the Aristotelian confusion of definitions with causes and essences with contingent properties effectively cut off the view that in science we ought to be listening to nature pushing back against our hypotheses and conjectures. On the contrary, the prevailing view in antiquity was that science could proceed largely a priori in the sense that science was a question of getting one's fundamental definitions and prior metaphysics in order. For the ancient philosophers and scientists, the structure of nature was a function of our familiar categories of thought, not the other way around (as many now think who opt for what is called **naturalized epistemology**); and familiar categories of thought have a tendency to impose on the phenomenon under investigation what *seem* to be necessary, if not definitional, relations and connections among its structural aspects, relations and connections that in reality derive from unreflective (or even idiomatic) habits of speech erroneously read back into nature as though those habits represented objectively fixed necessities and eternal essences. That is one reason, among many, that the notion of explanation in science remained a neglected backwater in post-Aristotelian philosophy of science for nearly 2,000 years.

The first philosopher of science to begin loosening the stranglehold of apriorism was Francis Bacon (1561–1626), an English philosopher who between his more scholarly researches was involved in crown and international politics. Bacon's philosophy of science is nowadays generally disrespected for its alleged naivete: Bacon is charged with having adopted an oversimple conception of scientific method in which the practitioner becomes a mere cataloger of trivial observable differences among members of different classes of phenomena. I believe this to be a caricature of Bacon's model of science, but to pursue the debate about the matter here would take us too far offtrack. What is of direct interest to us at present is Bacon's open hostility to Aristotelian metaphysics and his insistence that, in general, scientific questions are settled by observational experimentation, not a priori reasoning. Bacon was a doctrinaire inductivist, and as such he was hostile to the deductivist tradition in philosophy of science.

Not for him were the deductivist scientific systems of Aristotle and Euclid. The problem this created for Baconian science was that it suffered from a lack of predictive power. Without a starting clutch of general principles, predicting the behavior of a system was sheer whimsical guesswork. Not just anything can happen in a natural system, although Baconians, at times, seemed to think the opposite.

Yet the natural universe *is* a constrained system of events, objects, and properties. Pine trees do not speak Latin, pure water does not solidify at 300 degrees Celsius, heavy exercise does not lower fever, and falling bodies do not accelerate proportional to their masses. Some possibilities cannot become actualities in the course of natural history. In other words, nature runs according to relatively stable regularities. Unfortunately for subsequent philosophy of science, the religious domination of culture during the Renaissance suggested to philosophers of those generations that the appropriate way to conceptualize these natural regularities was to understand them as laws of nature. I say that this was unfortunate because the notion of a law of nature derives ultimately from purely theological doctrines about the natural world—doctrines that have no proper place in a rigorously empirical view of nature. A *law* requires a lawmaker, of course, and so it was easy to slip into the view that the laws of nature were conceived and ordained by a supreme deity. This made the new science acceptable to the church on the grounds that the scientists were not really undermining piety with their researches, they were merely ferreting out the minute microstructural proclivities of God. But, as subsequent generations of philosophers have come to appreciate, the use of God in scientific explanations is vacuous and otiose. Appeals to the will of a deity cut off prediction, stifle unification of distinct phenomena under a common set of regularities, and render the explained phenomenon ultimately groundless (because its intrinsic features are ultimately due to the inscrutable choices of a deity).

Nevertheless, the notion of a law of nature took hold deep within philosophy of science. It is still very much with us today, to the consternation of a growing number of philosophers of science. Its contemporary popularity is due largely to the work of one of the twentieth century's most influential philosophers of science, Carl Hempel. Hempel and Paul Oppenheim published a paper in 1948 in which they presented an extraordinarily precise and fertile model of scientific explanation, a model that made essential use of the notion of laws of nature. Indeed, the subject of scientific explanation in its modern form dates from the publication of this paper. The model's dominance has been remarkable, to the point that alternative conceptions of scientific explanation are presented and evaluated in light of how they are different from Hempel and Oppenheim's model. A consensus is growing that the model is a hopelessly utopian one, nowhere exemplified except in a few airy reaches of theoretical physics. But its far-flung influence requires that we understand it thoroughly before declaring it dead.

6.1 The Deductive-Nomological (D-N) Model of Scientific Explanation

Hempel and Oppenheim began their classic paper by presenting an explanation of why the column of mercury in a mercury thermometer experiences a short drop before rising after the thermometer is suddenly placed in hot water. The heat from the water at first affects only the glass casing of the thermometer, causing the glass to expand and provide a greater volume for the mercury inside to occupy. The mercury therefore drops in height until such time as the conduction of heat from the glass to the mercury produces an expansion of the latter. Different substances expand under heat at different rates and to different degrees, which the physicists represent by mathematical regularities involving the substances' various "coefficients of expansion." Mercury's coefficient of expansion is much larger than that of glass, and so the column of mercury eventually rises in the casing past the original starting point from which it at first dropped. What struck Hempel and Oppenheim about this explanation was its use of laws of nature, in this case certain laws of thermodynamics about heat flow and coefficients of expansion for different material substances. The event to be explained, the temporary drop in volume of the mercury encased in the glass tube of the thermometer, is made to appear a necessary consequence of the operation of certain regularities of nature, together with certain contingent facts about the thermometer in question. The event to be explained is "subsumed" under the relevant laws of nature, that is, it is "brought under" those laws in the sense of being a necessary consequence of their operation in the system during the time in question. Now, suppose that we have some way of linguistically representing all these relevant laws of nature, some way of expressing the laws in a fairly precise semantic form. Then, it ought to be possible to deduce logically a statement describing the event to be explained from a set of premises consisting of the laws of nature in question and further statements specifying the initial starting conditions, the contingent facts spoken of earlier. This basic idea is the core of the Hempelian model of scientific explanation, a model that Hempel and Oppenheim dubbed the **deductive-nomological model** of scientific explanation. The name is not difficult to understand. Explanation requires that a description of the event to be explained be *deducible* from a premise set containing the laws of nature under which that event is subsumable. Now the Greek word for 'law' is '*nomos*', hence it would be appropriate to name a model of scientific explanation that required deduction from laws a deductive-nomological model of scientific explanation. To save time and space, I will follow standard practice and shorten the name to the D-N model.

Hempel and Oppenheim recognized that not simply any old deduction-from-laws relationship will do, scientifically speaking; so, they placed four adequacy conditions in the D-N model, conditions that must be met in order to have a legitimate scientific explanation in hand.

First, the premise set consisting of the relevant laws of nature plus initial-condition statements must logically imply the description of the event to be explained. It is not enough that the description of the event to be explained is merely logically compatible with the premise set, that no contradiction follows from combining them. The description of the event to be explained must be a deductive consequence of the premise set under the standard inference rules of first-order logic.

Second, the premise set from which the description of the event to be explained is deducible must contain at least one law of nature, without which the description of the event to be explained is not deducible. It is not enough that the description of the event to be explained is deducible from other statements, none of which represents a general regularity in the universe. This is a subtle but important requirement of the D-N model, for it rules out as genuinely explanatory what are sometimes called narrative explanations, explanations that consist of telling a narrative story in which the event to be explained is simply placed within a sequence of other particular events. Hempel denied that such narrative stories are explanatory, because he argued that it is absurd to attribute determinative powers to mere places and times—but this is what a pure narrative must do, if it is taken as explanatory of some one event in the narrative. If you are told that first event *a* happened, then event *b* happened, then event *c* happened, then event *d* happened; and, furthermore, this narrative of itself explains the occurrence of event *c*, then you would seem to be forced to assume that *c*'s occurrence is due determinatively to the mere fact that it happened between events *b* and *d*, that it occurred when and where it did simply because it took place when and where it did. But, argued Hempel, this is superstitious nonsense. Mere places and times do not have determinative powers. What does? Laws of nature do. If, contrary to the mere narrative above, you are told that some property constitutive of event *b* is *nomologically sufficient* for some property constitutive of event *c*, now you've got a determinatively sound basis on which to explain the occurrence of event *c*—or so Hempel argued.

The third adequacy condition of the D-N model is a bit more arcane. Hempel and Oppenheim required that the statements composing the premise set from which the description of the event to be explained is deduced must be empirically testable. There must be a way or ways to test the premises independently of their use in the explanation in question, and this way of testing must not be merely conceptual (for example, not through a thought-experiment). The philosophical motivation behind this third adequacy condition involved certain doctrinaire views that descended from logical positivism. The goal was to rule out as legitimate science various traditional systems of thought like astrology and phrenology on the grounds that the "explanations" such pseudosciences provide rely on nonempirically meaningful principles and hypotheses, principles and hypotheses whose only evidential support is the dead weight of tradition and superstition.

The fourth adequacy condition of the D-N model underwent change. Initially Hempel and Oppenheim required that all the laws and initial-condition statements from which the description of the event to be explained was deduced be true laws and initial-condition statements. This was later weakened to the requirement that they only be well confirmed, because we clearly want to allow that there could be an explanatory relation between a description of some event and what are technically false laws, as well as between some event description and laws that cannot be shown to be true due to limitations on human epistemological capacities.

Suppose that we have two equal-sized groups of genetically identical mice. We inject all the members of one of the groups with a drug that inactivates interferons, but we leave all the members of the other group drug-free. Interferons are a class of cytokines, immunologically active biochemical molecules, secreted by certain classes of white blood cells in the immune systems of mammals. Interferons destroy viruses through a variety of different mechanisms, from activating other classes of immune cells that kill viruses directly to inducing the expression of antiviral defense mechanisms on bystander cells nearby, making the bystander cells impervious to infection by free virus particles. The capacity of interferons to do this is a lawlike feature of their chemical structures. Suppose that subsequent to injecting the first group of mice with the interferon-inactivating drug we expose both groups of mice to increasing concentrations of infectious virus particles, starting with small, subinfectious concentrations and increasing up to whatever concentration is required to kill 50 percent of the mice not injected with the interferon-inactivating drug. If we perform this procedure correctly, then we will observe that all the mice in the group injected with the interferon-inactivating drug die from viral infection at concentrations several hundred times less than the concentrations required to kill half the mice in the group not injected with the drug. Now an explanation of this experimental observation is not hard to construct. If we want to explain why any randomly chosen mouse, say Zeke, in the group given the drug died after exposure to a concentration, call it concentration *c*, of virus that was nonlethal to all the members of the other group of mice, we need only construct a deductive argument in which the premises contain the appropriate immunological laws about interferons and their causal effects plus statements indicating the relevant initial conditions: in this case, that Zeke was given a certain dose of the drug in question and that Zeke was exposed to such-and-such a concentration of a species of infectious virus. The statement that Zeke died a short time later is now deducible from the above premises by an application of the standard inference rules of first-order logic, universal instantiation, conjunction, and *modus ponens*. The argument in structural outline follows.

1. For any mammal x, if x is given an interferon-inactivating drug and x is exposed to infectious virus particles above concentration *c*, then x dies.

2. Zeke was given an interferon-inactivating drug.
3. Zeke was exposed to infectious virus particles above concentration *c*.
4. If Zeke is given an interferon-inactivating drug and Zeke is exposed to infectious virus particles above concentration *c*, then Zeke dies [1, UI].
5. Zeke was given an interferon-inactivating drug and Zeke was exposed to infectious virus particles above concentration *c* [2,3, Conj].
6. Zeke died [4,5 MP].

Premise 1 is a law of nature. Premises 2 and 3 are initial-condition statements. Premise 4 is deduced from premise 1 by the logical inference rule of universal instantiation. Premise 5 is deduced from premises 2 and 3 by the logical inference rule of conjunction. The conclusion 6 is deduced from premises 4 and 5 by the logical inference rule of *modus ponens.*

Notice also that this D-N explanation meets the four conditions of adequacy. The conclusion is deducible from the premises, it is not deducible from them without the law of nature represented by premise 1, each premise is empirically testable, and each premise is true or can be rendered well confirmed.

Admittedly there is something rather philosophically appealing about a nomic subsumption model of scientific explanation. It removes mystery and subjectivity from what would otherwise most likely be a mysterious and highly subjective judgment of explanatory fitness and power. Subsumption under law demystifies the determinative grounds of the phenomenon explained at the same time that it ensures a criterion of explanatory power independent of the subjective impressions of the scientific practitioner. Zeke's death is neither determinatively groundless nor is its explanation an impressionistic judgment call of the research immunologist. In Figure 6.1 I present another illustration of the D-N model of scientific explanation from the one domain of science in which the model works best—physics.

Unfortunately, every precise and tidy thing purchases its precision and tidiness at a significant cost, and the D-N model of scientific explanation is no exception. Two major sets of problems confront the model with what many philosophers of science consider to be flaws that are fatal to it. First, the model helps itself to the thorny and vexatious notion of laws of nature. Are there such things? If so, what are they, and how do we identify them as laws of nature? Second, even supposing we grant to the D-N devotee an unproblematic notion of laws of nature, the D-N model can be trivialized rather easily. That is, we can produce counterexamples to it, derivations that meet all the requirements and adequacy conditions of the model but which clearly are not explanations on other grounds. Such counterexamples, if genuine, indicate that the model is terribly incomplete, if not completely wide of its target. We will begin with the first set of problems, for without a clear notion of laws of nature the D-N model would have no special interest for us as a rich and fruitful account of scientific explanation.

The spectrum of a distant galaxy is produced by passing its light through a device called a spectrogaph, which breaks the light into its component wavelengths. In the resulting spectrum the galaxy appears as a bright smeared out blob of light running from shorter to longer wavelengths. In many cases dark vertical lines appear in the resulting spectrum. These are absorption lines. They are caused by the absorption of energy emitted by the component stars of the galaxy as the energy passes through cooler gases in the galaxy. Specific chemical elements in the cooler gas regions absorb the energy at specific wavelengths, thus removing that energy from the spectrum, leaving a dark space or line where the absorption occurred. Spectra are taken in earth laboratories to establish the standard positions of the absorption lines for each element. Hydrogen, for example, is associated with many spectral lines, including an absorption line at 487 nanometers (nm). Suppose the spectrum of a galaxy shows that the hydrogen absorption line normally found at 487 nm is instead found at 497 nm.

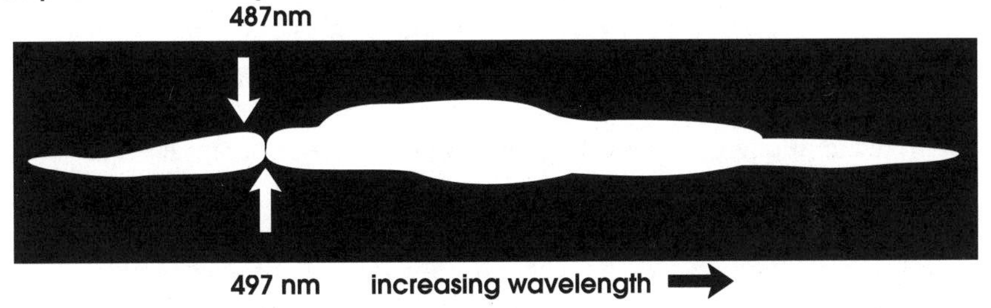

We may provide a D-N explanation of the 10 nm difference in wavelength by subsuming it under the relevant laws of physics and the relevant initial conditions. In this case the relevant laws of physics are,

1. **Hubble's Law for the expansion of the universe: $v=d \times h$**
2. **The constancy of the speed of light in a vacuum: $c=2.997 \times 10^8$ meters per second**
3. **The redshift law: $z=[(1+v/c)/(1-(v/c)^2)^{1/2}]-1$**

where v is the velocity of recession relative to the earth, d is the distance from the earth, and h is Hubble's parameter, which gives the current rate of expansion per unit distance. We know the value of h to a reasonable degree of confidence: 20 kilometers per second per 1,000,000 light-years (i.e. per 9.46×10^{21} meters). That value would appear as an initial condition premise. Another initial condition premise would be a statement of the standard laboratory wavelength for the spectral line of hydrogen at 487 nm. We include as a final initial condition premise a statement of the distance to the galaxy in question. There are a number of different methods for measuring distance to galaxies. Let us suppose in our example that using the cepheid variable method we have arrived at a measure of the distance to the galaxy in question: d=310,000,000 light-years (i.e. 2.933×10^{24} meters). We may now mathematically derive v for the galaxy using Hubble's law: v=6,201 kilometers per second. Using v we may mathematically derive the galaxy's redshift using laws 2 and 3 above: z=0.021. Finally, multiplying the standard laboratory wavelength 487 nm by z will give us the difference in observed wavelength: $z \times 487$ nm=10.18 nm, i.e. 10nm.

Figure 6.1. A D-N explanation of the redshift of galactic spectral lines.

6.1.1 Laws of Nature

The vast majority of contemporary philosophers of science do not read any theological connotations into the notion of a law of nature. For them lawlikeness has lost its original element of divine ordination; and rightly so, for determinative power is now understood in many cases to be a function of structure and not divine will. One might therefore ask, why do philosophers of science continue to use the notion of laws of nature when the lawgiver has dropped out of the picture? What does one gain by adopting the language of lawlikeness without any sovereign in the background whose very existence would otherwise seem to be a precondition of lawlikeness? Those are not easy questions to answer in a short and convenient way. I have already indicated one reason why Hempel insisted on the presence of deductively necessary laws in the premises of a D-N explanation: Only subsumption under law can avoid the attribution of determinative powers to mere places and times. But that reason doesn't really address the present issue, for we can still insist on an account of how the laws remain lawlike minus any divine lawgiver to sustain them by force of will.

One answer to the question is to claim like the philosopher David Hume that the fundamental laws of nature are "brute," without more elementary grounds, the naked verities we must accept despite their seeming arbitrariness. *They* are what explain everything else, everything less fundamental, but they themselves have no explanation in terms of something more basic to the universe (not even the divine fiat of a deity). What contemporary philosophers of science like about laws of nature is not so much their metaphysical pedigree (such as the fact that we used to attribute their being what they are to a deity) but what we might call their logical properties: their universal scope of application and their subjunctive import.

Laws of nature are not system (that is, spatiotemporally) specific. There is no law of nature that is solely about the planet Jupiter, or solely about oxygen molecules within the city limits of Denver, Colorado. If there is some microstructural aspect of oxygen molecules that is lawlike stable, *all* oxygen molecules *everywhere* manifest that aspect, and therefore they all fall under the law of nature in question. If there is some physical feature of the planet Jupiter which is lawlike stable, *all* planets *anywhere* which possess the same physical feature fall under the same law. Let us call this the *universal scope* of laws of nature. This terminology comes from the positivists, who of course wished to capture lawlikeness linguistically, if at all possible. The universal quantifier in standard first-order logic is the obvious tool to use here, combined with the material conditional ('if, then') connective. That was the received analysis of laws of nature under the positivist model of scientific theories: laws are universally quantified material conditionals. We made use of this in constructing the dummy theory to illustrate the positivist model in section 2.2, and the reader may wish at this point to review briefly that section.

The problem with treating laws of nature as universally quantified material conditionals is that there are universally quantified material conditionals that are true but nevertheless not laws of nature; hence, the proposed analysis does not specify a sufficient condition for lawlikeness.

> For all objects x, if x is a moon of the earth, then x has less than 1 percent oxygen in its atmosphere

is a true universally quantified material conditional, but it is not a law of nature. It is, we might say, an "accidental" truth that the earth only has one moon and that the moon has less than 1 percent oxygen in its atmosphere; for if certain physically possible astronomical events had occurred in the remote past, or else other physically possible astronomical events were to occur in the future, then the conditional would have been or will be false. Its truth is too vulnerable to being overturned by contingent events. A different history to the formation of the solar system or a future gravitational capturing of a passing large astronomical body by the earth's gravitational field could each render the conditional false. But a law of nature, we want to say, could not possibly be false, and it is this *necessity* to lawlikeness that is missing from the simple positivist analysis.

Unfortunately, the positivists had a pet peeve against *modal* concepts like possibility and necessity, so they refused to amend their analysis by "going modal" openly. Instead, they sought to tinker with the original analysis, somehow upgrading it into a more workable version. Hempel suggested at one point that what was wrong with accidently true universally quantified material conditionals (like the one about the moon's atmosphere) was that they mentioned particular objects, places, times, or events. So, he insisted that the predicates in a law of nature must be purely qualitative predicates, where a purely qualitative predicate is one whose definition, or statement of *meaning*, does not make reference to particular objects, places, times, or events. This requirement struck many philosophers of science as no improvement whatsoever. Many critics claimed to see no superiority to an account of lawlikeness that used the notion of mysterious things called meanings over an account which simply went modal and stuck 'necessarily' (what modal logicians call a necessity operator) in the front of the conditional. How is a meaning a more understandable feature of the world, a more natural aspect of reality, which makes it an acceptable notion to use in constructing a model of scientific explanation, whereas the use of a formal necessity operator from model logic would be supposedly unacceptable? The answer would seem to be that there are no grounds of ontological parsimony for preferring meanings to necessities—they are each intractably semantic notions. Accordingly, most contemporary philosophers of science are perfectly happy to accord laws of nature a form of necessity (usually called physical necessity). One objection often made against such a cavalierly modal version of lawlikeness is that it is circular: for physical necessity itself is ultimately a matter of what the laws of nature are in a given possible world. The idea is that what *must* occur in the course of the world's history is a function of what the fundamental laws in the world ordain. But the correct response to this objection is to point out that this is a circle even the positivist cannot get out of—this is a circle in which we are trapped. What must happen is a function of the fundamental laws of nature, and those laws are fundamental because they underlie everything else and nothing underlies them. We are back to Humean bruteness, again. I do not find that an unhappy place to land, for I'm inclined to think that it was illusory to suppose we could ever give a noncircular account of what makes the fundamental laws of nature laws of nature. They just are, and one must come to feel comfortable with that state of affairs.

Further, the positivist criterion for lawlikeness may not even provide a necessary condition for lawlikeness; for there may be laws of nature that are species specific, that is, laws that essentially are about one particular species of biological organisms (and so such laws violate the stricture against mentioning particular objects, places or times). For example, it may be that certain psychological properties of higher organisms are lawlike connected to entirely different families of neurophysiological properties depending on the species to which the organisms respectively belong. Human pain may not be physiologically the same as canine pain. Nevertheless, it may be the case that all canine pain is lawlike determined by one family of canine neurophysiological properties. Here would be a species specific law of nature. This notion of nonuniversal laws is controversial, and I do not wish to defend it in depth here. I do not rule it out on conceptual grounds alone, however, as some positivists no doubt would. I merely want to note that if the evidence should turn out to support such a species specific conception of laws of nature, then the requirement that laws of nature be universally quantified conditionals with purely qualitative predicates would not even provide a necessary condition for lawlikeness.

6.1.2 Standard Counterexamples to the D-N Model

It is traditional in a philosophy of science text to present a host of standard counterexamples to the D-N model of scientific explanation: either cases that fit the model but are not explanations on other grounds, or cases that don't fit the model but are explanations on other grounds. This can be a tedious exercise, and it often leaves students with a sense of irritation at the technical trickery in whose service formal logic can be put. I don't want to belabor these counterexamples, but I do want to make the student aware of the larger point they make, a point that often gets lost amid all the technical logical wizardry. Here is the point: Yet again, the positivist attempt to model an important objective relationship with the use of symbolic logic hits the rocks and splinters into pieces. What important objective relationship? Determination. For it is one of the theses of this book that *there is no explanation without determination.* If we have an explanatory relation between a premise-set and a description of the event to be explained, then that can be only because something in the premise-set adequately captures the objective determinative connection between the events in question. The failure of the D-N model of scientific explanation has do to with the inability of mathematical logic *alone* to represent objective determinative connections; for all logical connectives represent are syntactic and semantic connections, not causal or supervenient ones.

It is no wonder then that sufficiency counterexamples to the D-N model can be had by the dozen. These are cases where all the formal conditions of the D-

N model are met, but there clearly is no explanatory relation between anything in the premise-set and the description of the event allegedly explained. Perhaps the most famous of these sufficiency counterexamples is the flagpole case. Attached to the top of a vertical flag pole is a 50-foot stabilization wire, which runs taut to the ground where it is attached 40 feet from the base of the flagpole. Elementary geometry, in particular the Pythagorean theorem (a mathematical law) that says that the square of the hypotenuse of a right triangle is equal to the sum of the squares of the other two sides, together with the initial conditions about the length of the wire and its length from the base of the pole, would allow us to deduce that the flagpole is 30 feet high. All the adequacy conditions of the D-N model are met in this case, but surely the explanation for why the flagpole is 30 feet high has nothing to do with the stabilization wire and the Pythagorean theorem. Presumably, the real causation of the pole's height (the real explanation of its height) is a messy matter of bureaucratic public policy. Some bureaucrat in the Department of Public Works decided that the pole would be 30 feet high after choosing from the menu of heights manufactured. Clearly, we don't have the determinatively relevant law in this case; but nothing in the D-N model works to ensure determinative relevancy. Hempel allocated that issue to the backwater of "explanatory pragmatics" when it ought to be the primary focus of any model of explanation. In a slightly more pathetic version of this same problem, consider the following D-N "explanation" of why you ate for dinner yesterday what you in fact ate. Let 'E' denote what you ate for dinner yesterday. Let 'L' denote the law: all activated macrophages produce interleukin-1. By the use of the valid deductive inference pattern known as disjunctive syllogism, we may D-N explain why you ate what you did for dinner yesterday:

1. E or not L
2. L

therefore 3. E

Clearly, we need a richer model of scientific explanation, one that takes causation and determination in general very seriously and right up-front. We shall take a look at one such model in the next section.

6.2 The Aleatory Model of Scientific Explanation

The philosopher of science Paul Humphreys presented a thorough model of scientific explanation in his book *The Chances of Explanation*, a model in which he sought to develop an account of causal explanation geared primarily toward probabilistic natural processes.

Humphreys calls his model of scientific explanation the **aleatory model**, from 'alea', a Latin name of a game of dice or chance. The basic idea is that an explanation cites a cause of the event being explained, and a cause is something that contributes invariantly to an alteration in the probability that its effect will occur. Schematically, an event C causally influences event E if C alters the probability that E will occur *invariantly*. The key novelty here is Humphreys' use of invariance. This is a technical notion in causal modeling theory, which Humphreys gives a rigorous formal treatment; but, for our purposes, the idea can be simplified. A phenomenon is invariant when it occurs despite alterations in other background conditions—if, no matter what else you change in the situation, the phenomenon occurs nonetheless. Causes invariantly modify the chance of occurrence of their effects. Some such causes modify the chance of the effect's occurring to the negative (they lower the likelihood), and Humphreys calls such causes counteracting causes; others contribute positively to the chance of the effect's occurring, and Humphreys calls those contributing causes. Scientific explanations are therefore mostly contrastive explanations: They tell you that C caused E despite counteracting cause A. An explanation usually takes place in a larger context in which the foreground contributing cause's potential defeat by a background counteracting cause is acknowledged.

Humphreys takes probability as his basic notion. Causation, and ultimately explanation, are accounted for in terms of probability. A cause is any factor that invariantly modifies the probability of occurrence of an effect, and explanations cite causes. The invariance condition Humphreys imposes is quite strict: The reputed causal factor must modify the probability of the effect in the same way no matter how we alter all other circumstances consistently with the causal factor's occurring at all. Humphreys points out that the exact value of the probability change is not relevant, only the fact that it is an increase or a decrease from the prior probability of the effect's happening. So, to take an example, it is permissible to cite the physiological work of X's NK cells in explaining the remission of X's cancerous tumor, only if the work of such cells increases the probability of tumor remission *in all circumstances consistent with the presence of NK cells and tumors of the given cell type*. If we alter X's body chemistry in any way short of a way that makes it impossible for either NK cells or the given type of cancer cells to exist in X's body, and the work of the NK cells still increases the probability of tumor remission, then the work of such cells is a cause of the tumor's remission, despite the presence of countervailing factors such as the various proteins and enzymes that cancer cells produce in an effort to nullify the effects of NK cells. Or again, the presence of specific E-class antibodies to weed pollen in the respiratory mucous membranes of X may be cited in explaining X's allergic symptoms only if the presence of such E-class antibodies increases the probability of X's experiencing allergic symptoms *in all circumstances consistent with the presence of such antibodies and weed pollens in the respiratory membranes of X*.

The main problem with the aleatory model of scientific explanation, discussed at length by the philosopher of science James Woodward, is that it demands that no factor counts as a cause capable of explaining its effect unless its modification of the probability of the effect is invariant across *all* possible circumstances consistent with the occurrence of the factor and its effect. This is too strong, for it requires that virtually *every* single minute condition that *might* effect the factor's modification of the probability must be taken into account before the factor can be declared a cause of the effect in question. This is surely too utopian, even for something as idealistic as theoretical science. Under such a requirement we cannot declare NK cells to be causes of tumor remissions unless we first are sure that we have considered *all possible* interfering circumstances consistent with the existence of NK cells and tumor remissions. This seems rather unreasonably strict, and insofar as the aleatory model demands total inclusion of all possible relevant interfering factors, it is too fine-grained. Under its requirements, the only causes we shall ever discover are "partial" causes (for we haven't confirmed the invariance against all possible circumstances), and the only explanations we shall ever have are "partial" explanations. Some philosophers might find such a view appealing, because they are fond of anything that reminds us of our epistemological humility in the face of a microstructurally complex universe; but, in general, such a view cuts against the commonsense truth that mature sciences do explain many events in their domains of inquiry, and that they do so in a relatively complete sense of 'explain', boundary conditions of the system in question held fixed, of course.

Something that seems to me to be a definitely positive feature of the aleatory model of scientific explanation, at least as Humphreys construes it, is that it places explanation on a more honest ontological basis than does the D-N model. Humphreys states that, ultimately, a structural account of why and how causes modify probabilities is to be sought. The basis for why and how causal factors modify probabilities is grounded in the microstructural changes that such causal factors produce in the systems in which they operate. This seems to me to be completely on the right track, for explanation comes to an end with an account of microstructure. Hence, the real determinative force that underlies an aleatory explanation is not provided by what the probabilities are doing (increasing or decreasing), but in the presumed microstructural changes that the alterations in probabilities are the detectable results of. Humphreys seems to assume that the microstructural changes are not simply a matter of more alterations of more complex probabilities; otherwise, cashing out the probability alterations in terms of microstructural changes would purchase no explanatory gain, and certainly no ultimate grounding of explanation. One senses that the aleatory model therefore fails to address itself to the "real action" taking place at the microscale of nature. Let us look at yet another alternative model of scientific explanation that has become influential in very recent times.

6.3 The Unification Model of Scientific Explanation

The philosopher of science Philip Kitcher believes that it is a mistake to evaluate the scientific worth of explanations one at a time. One cannot tell if any given individual explanation is scientifically acceptable except in relation to how that given explanation fits into a larger family of explanations; yet the D-N and aleatory models presuppose that individual explanations stand or fall on their own. Kitcher thinks he has a better idea. Take all the beliefs that a given community of practitioners hold at any one time, call that body of beliefs K. Assume that K is representable linguistically: It is one large set of statements accepted as true by the community of practitioners. Now, the metaphysical point of science is to demystify nature, to render as small as possible the number of states of affairs we must take as brute, to render as large as possible the number of states of affairs that we can treat as derivative from more basic (brute) states of affairs. So, what matters to scientific practitioners is how the beliefs in K are systematized, how they are related to each other in a hierarchy of basic versus derived. For example, suppose K consists of 100 statements taken as true. There will be different ways in which those 100 statements can be hierarchically organized. Perhaps only three of them are brute and the other ninety-seven can be derived from those three using four elementary logical inference patterns. Or perhaps seven of them are basic and the remaining ninety-three are derivable from those seven using six elementary logical inference patterns. To Kitcher, the first way of organizing the set of 100 statements is, other things being equal (which we will discuss shortly), a more *unified* way of organizing them than is the second way. Why, you ask? Because, in the first way, a larger number of diverse factual truths are logically derivable using a smaller number of basic truths and a smaller number of basic inference patterns. To explain is to unify diverse phenomena by deriving their descriptions using the same small number of argument patterns over and over.

Notice certain similarities between Kitcher's approach and the classical D-N model. Both take explanation to require logical derivation of some description of the event to be explained. Both models therefore assume that an explanation is a logical argument of some kind. Kitcher goes so far as to embrace what he calls "deductive chauvinism," the view that all explanations are ultimately deductive, none inductive (so he must give a rather Byzantine account of probabilistic explanation in a field like quantum physics). Not even Hempel went that far in his wildest positivist mood. Both models reject direct appeal to causation, in contrast to the aleatory model. Both models instead take it for granted that the pattern of argument embodied in an explanatory derivation contains important information about the determinative structure of the event explained. This is perhaps more obvious in Kitcher's case than for the D-N model. To see why this is so requires that we learn a few details of Kitcher's **unification model** of scientific explanation.

The basic idea of the unification model is simple when stated in slogan form, but rather complicated when actually carried out: A derivation of the description of an event is an explanation of that event provided that the pattern of argument represented by that derivation is a member of what Kitcher dubs the *explanatory store* of the community of practitioners in question. The explanatory store is that collection of argument *patterns* (not arguments themselves) that best unifies the body of beliefs currently held by the community of practitioners in question. A collection of argument patterns P best unifies a body of accepted statements K if three conditions are fulfilled:

(i) No other alternative set of argument patterns that imply the same set of accepted statements K contains a smaller number of distinct patterns than does P.
(ii) No other alternative set of argument patterns that contains the same or a smaller number of distinct patterns than P implies a set of accepted statements larger than K.
(iii) Any alternative set of argument patterns with the same number of distinct patterns as P and that implies the same accepted statements as K contains argument patterns that are less stringent than the patterns in P.

This is all a bit hazy and technically esoteric. Condition (iii) is surely opaque to anyone not trained in formal logic; for the notion of stringency is a largely syntactic one, having on one version to do with the number of different nonlogical symbols in a symbolic formula one may permissibly substitute for. The symbolic formula 'Px v Py' is more stringent than the symbolic formula 'Γx v Py', for in the former only the 'x' and 'y' variables over individuals may be substituted for, while in the latter the individual variables 'x' and 'y' and the predicate variable 'Γ' may be substituted for. Why does Kitcher foist upon his unification model of scientific explanation such cumbersome and exotic logical complexity? The answer is instructive. It is because his test of unification is quantitative (as, indeed, presumably any test of unification would be, given that it is an intrinsically quantitative concept). We must be able to *count* argument patterns, accepted truths, and substitution places in symbolic formulae. Without the ability to quantify such things, no content can be given to the idea that the argument patterns in one set of argument patterns logically imply *more* accepted truths using *fewer* basic patterns than the argument patterns in any other set of *equally or more* stringent argument patterns applied to the same body of accepted truths. This requirement that we be able to count such things leaves the model open to technical problems of which Kitcher is well aware. Once again, the attempt to represent the causal determination that underwrites explanation by purely logical relations threatens to derail what is otherwise a basically sound idea: Explain as much of the unfamiliar using as little of the familiar as you possibly can get away with. An example will help us to see the intuitive appeal of this idea.

Once it became known to immunologists that the blood of a mammal is loaded with protein antibodies that each have a specificity for binding to only a few antigens, the question naturally arose of how to explain this extremely narrow specificity. It is not as though there are but a few generic antibodies and each of them fits thousands of chemically distinct antigens. The antibody to a particular species of weed pollen, for example, binds basically only to that species of weed pollen, or to just a few other species of weed pollen very closely related genetically to the given species. Over the course of forty years, between 1920 and 1960, two competing accounts were proposed to explain the narrow specificity of antibodies. The antigen-template theory held that the antigen, the substance the antibody will bind, acts as a kind of physical mold (like a cookie cutter) to which the antibody conforms its shape so as to be able to bind to it. This theory did explain certain aspects of the specificity of antibodies, most notably how there could be such a large number of distinctly specific antibodies (they are made on the spot by template conformation to the antigen). But there were a host of problems associated with the antigen-template theory. It was difficult to account for the persistence of antibodies in the absence of antigen, and there seemed no obvious way that the theory could be used to explain anamnestic boosting, the phenomenon whereby a tiny second dose of an antigen elicits an enormous antibody response, a response bigger than the original response to the initial larger sensitizing dose of the antigen.

Perhaps the most serious problem with the antigen-template theory was that it postulated a form of chemical template molding that seemed strangely unique to the immune system. Why there should be such a class of moldable proteins in the immune systems of mammals was a biological oddity—it fit in with nothing else in the biochemical world. Many creative hypotheses were put forward to account for such template molding: enzyme modification, tertiary folding of preformed amino acid chains, indirect templates. None of these hypotheses succeeded in explaining the specificity of antibodies to the satisfaction of practicing immunologists. New evidence accumulated that damaged the theory's credibility even more. Repeated immunization with the same antigen results in *qualitative* changes in the antibodies formed, not merely changes in the quantity of the antibodies: The later antibodies may be more or less specific in their binding power than the earlier antibodies, but why that should be so given that it was still the same antigen template doing the molding of the antibodies was mysterious.

In the 1950s a change in orientation swept through immunology. The prior dominant crowd of physicochemically inclined theorists was replaced by a generation of biologically oriented immunologists. The leading figures were the future Nobel laureates F. Macfarlane Burnet and Niels Jerne. Between 1955 and 1958, along with David Talmage and Joshua Lederberg, they worked out the clonal selection theory, now the reigning model of antibody formation in contemporary immunology. The clonal selection theory accounts for the formation of antibodies with narrow specificity by positing the existence of an innate genetic repertoire of immune *cells* (not innate proteins) each one of which is genetically committed to the potential manufacture of one specific antibody. What a given antigen does is to act as a natural selector of those immune cells that happen to make an antibody chemically capable of binding to it, selecting those cells, and those cells only, from among the vast array of patrolling immune cells, causing them to differentiate and start churning out copies, lots of copies, of the one antibody they are capable of making. There is no template molding, only a process of miniature natural selection of immune cells that happen to manufacture appropriately fitting antibodies (by accident, we might say). Now here was a theory that could account for many of the facts that the antigen-template theory could not. Antibodies persist in the circulation in the absence of fresh antigen because the cells that make those antibodies have been selected and remain in the circulation for some time. The anamnestic response is due to the fact that, once sensitized by an initial dose of antigen, many immune cells capable of making the corresponding antibody are circulating throughout the body in a non-activated but "ready" state. On contact with even a small second dose of the same antigen an enormous explosive response ensues as all those cells are activated and begin producing antibodies. Further, the alteration in qualitative binding power of antibodies formed after repeated immunizations with the same antigen can be accounted for by means of selection of more specific or less specific clones.

I believe that this competition between the antigen-template theory and the clonal selection theory illustrates the basic idea behind Kitcher's unification model of scientific explanation. The antigen-template theory is an explanatory account less unifying than the clonal selection theory. The latter theory is simply a localized version of general natural selection—natural selection shrunk down to the size of an individual organism's immune system. The patterns of derivation, to use Kitcher's mode of characterization, used to generate explanations of immune phenomena under the clonal selection theory bear striking structural similarities to those patterns used in other fields of biological inquiry such as ethology, population genetics, comparative physiology, and so on. We have a natural source of genetic variance of traits (in this case, a genetic repertoire of antibody producing B cells), a source of selection pressure (processed antigen), and the differential activation of the selected trait source (activated clones that make the fitting antibody). We have increased the range of facts that the same argument patterns (which represent natural selection mechanisms) can be used to derive an account of. The same cannot be said for the antigen-template theory. There theorists had to work overtime trying to conjure up uniquely bizarre mechanisms for how the template molding took place at the microstructural level. If successful, any account of it would have been the very opposite of unifying; on the contrary, it would have counted as a new kind of biochemical process to be added to the stock of those we already knew about. In such a case, the unfamiliar would have been explained in terms of the equally unfamiliar, increasing rather than decreasing the number of physical processes we must take as causally independent and theoretically basic.

But we have not yet asked the critical questions that need to be asked of the unification model: Why should we prefer explanations that unify to those that do not unify? Why is it preferable to explain lots with little? Why should nature be such that it is the outcome of the effects of a small number of basic processes rather than of the effects of a large number of basic processes? Why couldn't the universe be, as Kitcher puts it, "messy"? If nature was messy, then the search for unifying explanations might generate inaccurate explanations. In such a case, what we think is going on causally under our highly unified theories is not what is really going on down there in the microstructural trenches of nature. Kitcher's defense of the unification model against this legitimate worry strikes this author as giving away the game at the last moment. Kitcher points out that the worry in question presupposes that the "causal order" of nature is independent of our theoretical systematization of nature—that causation is independent of theory. Only if one presupposes that the causal relations among events exist independently of what our unifying theories say they are will one worry that the unifying explanations sanctioned under those theories may get the causal facts wrong.

Kitcher denies that the causal order of the world is independent of our theoretical systematization of the world. He denies that there could be a cause C of some phenomenon P which was not referred to by some derivation in the explanatory store at the limit of scientific inquiry; for, he claims, no sense can be made of the notion that there could be a form of causal influence that is not represented by our completed scientific theories. Kitcher cites the philosophers Immanuel Kant and Charles Sanders Peirce as supporters of this view. In particular, Kitcher makes use of Peirce's notion of a limit of scientific inquiry that we were introduced to in chapter 4. Assuming that the human race evolves forward into the future indefinitely, we can imagine the situation where our scientific theories develop ever closer to perfection without actually obtaining it, where our total knowledge of nature reaches some stage of completion that approaches perfection without ever reaching perfection. The notion of an approach to a target that moves closer and closer to it, indefinitely, without ever actually reaching it, is similar to the purely mathematical concept of a limit, familiar to many readers, no doubt, from their calculus courses. So, imagine scientific inquiry proceeding through time to the limit, to that stage of maximal completion that it is capable of but which is short of perfection. This turns out to be a useful notion for discussing a variety of topics in philosophy of science, epistemology, and metaphysics. Here Kitcher makes use of it to argue for what will strike some people as a distressing conclusion: What counts as a cause will be a function of what our completed, maximally unified systematization of the world in the limit of inquiry says is a cause. Under the unification model of scientific explanation no causal mechanism can be explanatorily relevant that is not represented in our scientific theories in the limit of inquiry, for such representation in the limit of inquiry, Kitcher argues, is "constitutive" of explanatory relevance.

The unification model, writes Kitcher,

> . . . makes it constitutive of explanatory relevance that there be no causal mechanisms that are not captured in the limit of attempts to systematize our beliefs. ("Explanatory Unification and the Causal Structure of the World," p. 499)

What makes this a distressing claim is that it comes perilously close to being a form of antirealism, the view that the facts are partly determined by *our* activities of inquiry, that they don't obtain independently of whether or not we are around to discover them. The causal facts of the matter are "constituted" by our practices of inquiry—they are not "discovered" by our practices of inquiry. This is a position that is easier to swallow in some domains than in others. Most people can accept it with respect to certain of the so-called soft sciences, like economics and criminology; but it seems, frankly, downright bizarre when applied to something like physics or chemistry. For example, the biochemical causes that allow an NK cell to kill its target would seem to most people to be a central case of a reality that obtains independently of any human practices of inquiry, whatsoever. If NK cells kill by releasing perforin molecules that puncture holes in the cell membranes of target cells, then presumably they have done so for the millions of years since they evolved in mammalian immune systems, will continue to do so in the immediate future, and most likely always will do so (unless they become extinct at some future time), regardless of whether human science had ever even been invented in the first place. That is the position of the scientific realist, who adopts the view that it is possible for there to be facts of the matter (in this case, causal facts of the matter) that transcend all the evidence we have collected in the limit of scientific inquiry and which therefore are missing from representation in our maximally systematized theories in the limit of scientific inquiry. (Note: Whether their being missing from our theories in the limit of inquiry is necessary, or merely an accident of human imperfection, is a deeper dispute that we will postpone discussing until chapter 10.) Most practicing scientists, and quite a few philosophers of science, are realists in this sense. For them, there is not the slightest reason to suppose that our theories must capture 100 percent of reality, even in the limit of inquiry, or that what it will capture in the limit of inquiry must be captured with 100 percent accuracy. There may be evidence transcending facts, facts that transcend not only our actual evidence, but all possible evidence. These sorts of views are currently the subject of intense disputation in the philosophy of science, which is one reason that we shall give them their own chapter (see chapter 10).

Insofar as Kitcher believes that the best way to relieve the worry about the possible nonunified messiness of nature is to go whole hog and wed the unification model of scientific explanation to a form of antirealism, the model is, to my mind, severely weakened in its degree of plausibility. What started out as an interesting idea—explanation unifies the most with the least—has turned into an overtechnical and unintuitive account of science with metaphysically distasteful and doctrinaire consequences. If one has to adopt the view that causal relations are socially constructed rather than discovered as they are in themselves in order to advocate the unification model of scientific explanation, then so much the worse for the unification model. Such a view of science, as we shall see in the next chapter, may be currently the fashionable rage among trendy humanists and other intellectuals, but it hardly makes sense of the practices of the vast majority of workaday scientists. Perhaps in the future an argument alleviating the worry about a nonunified and messy nature may be produced that does not amount to simply denying a causal reality independent of our theories in the limit; but, until then, our verdict on the unification model of scientific explanation ought to be that it lets us down metaphysically just when we were getting ready to overlook its technical baroqueness and treat it as a serious candidate.

6.4 Natural Kinds, Homeostatic Property Clusters, and the Structural Basis of Reality

The last section ended on a metaphysical note, illustrating the fact that issues involving explanation bear on issues of ontology and metaphysics (and vice versa). The last section also illustrated the tension that exists between two general orientations within philosophy of science: the orientation that views the causal/determinative order of nature as independent of human practices of inquiry and explanation, and the orientation that views the two as permanently intertwined. Are the scientific facts discovered as they are in themselves, or are they at least partially constructed as they are in themselves by the idiosyncrasies of our evolving practices of inquiry and explanation?

The appeal of the classical D-N model was based on its requirement that the explanation of an event required subsuming it under the relevant laws of nature. Admittedly, the ontological status of those laws was always a problem for devotees of the D-N model. The metaphysical underpinning of lawlikeness was downplayed by Hempel, who chose instead to highlight the logical features of lawlikeness—the sufficiency of the antecedent cause for its consequent effect, the total generality of the law represented by the universal quantifier with which it was expressed in symbolic notation, and the subjunctive mood in which laws must be expressed. It was a largely unmentioned background assumption that the predicates in a law of nature denoted natural kinds—unmentioned because the times were still suffused enough with the spirit of positivism to make philosophers of science skittish about openly acknowledging metaphysical presuppositions. Most everybody would agree that, at least as an ideal, it would be a good thing if all the predicates in the laws of a given theory denoted natural kinds. Nobody objects per se to the notion of natural kinds as a wonderfully ideal concept, it is just that quite a few philosophers doubt if there are any such natural kinds in the sloppy and dirty real world that we inhabit. That is because the notion of a natural kind brings with it a whole tradition of ancient prejudices and related concepts that these philosophers object to (and perhaps rightfully so). Aristotle, who is usually credited with having worked out the first account of natural kinds, thought of them as abstract essences whose defining attributes were eternally fixed, unchanging, and absolutely precise. But nowadays we do not need to accept such questionable excess baggage to help ourselves to the concept of natural kinds. I think that what made the classical D-N model so appealing and gave it such lasting power is precisely that laws of nature do relate natural kinds, and that natural kinds represent the seams of nature, those structural divisions with which it simply comes to us in the course of our inquiry. But we have advanced the case for natural kinds far beyond what Aristotle was able to do. We no longer demand that natural kinds be eternally fixed, unchanging, and absolutely precise. In fact, they can be as sloppy as you'd like and still be useful in doing science.

The philosopher of science and realist Richard Boyd has recently worked out an interesting version of scientific realism that makes use of the concept of natural kinds, but only after the traditional notion of natural kinds has been reworked in a creative way. Kinds in nature, says Boyd, are not abstract essences (a la Aristotle), nor are they something mathematically esoteric that you can't understand unless you know formal logic (sets of possible worlds, for example). They are what Boyd dubs **homeostatic property clusters**, which we may call h.p.c.'s, for short. Boyd gives a thorough characterization of h.p.c.'s that is rather lengthy, but for our purposes the key features of h.p.c.'s are given by their name. They are clusters of properties that are homeostatically stable over significant intervals of time. The point of calling them clusters rather than sets is that their member properties fluctuate in a way that the members of a precisely defined set do not. The point of calling them homeostatically stable clusters is that the properties so clustered at any given time bear structural relations to each other such that changes in one are often compensated for by alterations in some or all of the others. The net effect is a kind of self-regulating stability, what in biological science is called homeostasis: The properties, while they individually come and go, nevertheless produce a stable cluster whose *causal powers* remain roughly constant over time. My guess is that the "underlying structures" Humphreys speculates are ultimately responsible for the alterations in probabilities that he claims constitute causation are homeostatic property clusters. I have no conceptual argument for this. It is an empirical speculation on my part. The important point to understand is that homeostasis is a structural concept—for that matter, so is serial clustering—for both have to do with the mechanics of constrained property covariance over time.

For example, it may well turn out to be the case that so-called NK cells form a natural kind, but only in the sense of being a homeostatic property cluster. To speak like a Platonist for a moment, NK-cellhood may not be a fixed and unchanging essence of any kind. NK cells may be intractably heterogeneous in their individual properties. Different subclasses of NK cells may kill through other mechanisms besides releasing perforin molecules. They may vary among each other in what surface proteins (what immunologists call surface markers and are identified using an internationally recognized naming system of "CD" [**c**luster of **d**ifferentiation] numbers) they bear on their surfaces. They are a species of immune effector cell subject to ongoing selection pressures within the biological life of the organism. Yet, as a class, their causal powers—what they can do causally, namely kill virally infected cells and some cancer cells—remain remarkably stable through all this microvariance.

What I want to suggest here should be familiar by now to the reader. I am a fan of appeals to structure. Such appeals end explanation like nothing else ends explanation. Nothing grounds determination like structure. Once you know the structural determination, you know the determination. Once you know the determination, you know the explanation. Accordingly, I am inclined to the view that any model of scientific explanation that does not cash out the explanatory relation ultimately in terms of structural aspects of the phenomenon explained—be it h.p.c.'s or something else—is traipsing down the wrong path. The problem this creates for the D-N model and for the unification model is obvious. Both models treat explanation as a primarily linguistic activity of some kind, as the making of certain kinds of structured formal derivations. Hence, both models will be successful only if the structural features of those formal derivations happily match the structural aspects of the external phenomena allegedly explained. But what rational grounds could there be for expecting to find such a happy coincidence between the order of our linguistic representations and the structural order of nature? Humphreys' model does not treat explanations as derivations. He at least acknowledges the prior structural basis of causal reality, and he knows enough to know that it is hopelessly utopian to expect a priori a structural match between the languages in which our theories are couched and the phenomena those theories purportedly deal with. Yet his model seems a tad toothless for its own reasons, given that it does not deal directly with these underlying structures that are responsible for altering the causal probabilities that the model is centered on. That causes alter probabilities seems to me to be something any rational person ought to grant. *How* they do so is just where all the details of real interest and power are located; yet that is precisely what Humphreys' model has nothing specific to say about.

These reflections bring to a close our discussion of the positivist age in philosophy of science, an age that saw the field develop to maturity through a series of powerful logical models of scientific concepts. But all ages must come to an end, for change is the only certainty of life. Philosophy of science underwent a postpositivist upheaval that began in the late 1950s and came to a head with a single publishing event in 1962 of such overwhelming impact and importance as to constitute a watershed in the philosophy of science, the publication of

Thomas Kuhn's masterpiece, *The Structure of Scientific Revolutions*. That work heralded a sea change in philosophy of science, a change in orientation, attitude, method, and substantive doctrine that we are still feeling the effects of at the present hour. To this recent philosophical upheaval we turn in the next chapter.

Further Readings

The literature on scientific explanation is vast but a bit technically tedious. The place to begin is with Carl Hempel and Paul Oppenheim, "Studies in the Logic of Explanation," first published in 1948 but reprinted in Hempel's collection of papers *Aspects of Scientific Explanation*, New York, Free Press, 1965, pp. 245–90. Wesley Salmon has written *the* definitive historical review covering all major work on the theory of scientific explanation since the original model proposed by Hempel and Oppenheim, "Four Decades of Scientific Explanation," in *Minnesota Studies in the Philosophy of Science Volume XIII*, edited by P. Kitcher and W. Salmon, Minneapolis, University of Minnesota Press, 1989, pp. 3–219. Salmon's review is now available as a separate work from the University of Minnesota Press. Paul Humphreys' aleatory model of scientific explanation receives its full treatment in his *The Chances of Explanation*, Princeton, Princeton University Press, 1989. Philip Kitcher discusses his unification model of explanation in "Explanatory Unification and the Causal Structure of the World," in *Minnesota Studies in the Philosophy of Science Volume XIII*, edited by P. Kitcher and W. Salmon, Minneapolis, University of Minnesota Press, 1990, pp. 410–506. Richard Boyd introduces the notion of homeostatic property clusters in his, "How to Be a Moral Realist," in *Essays on Moral Realism*, edited by G. Sayre-McCord, Ithaca, Cornell University Press, 1988, pp. 181–228.

7

The Revenge of Historicism

Intellectual upheavals are by nature exciting events. Often such an upheaval is sudden, having apparently come roaring out of nowhere, as the saying goes. In other cases there is a long prodromal period, a period of restlessness and rumbling during which subtle signs of incipient crisis are there for the observing, if only people were observant enough to detect them. After World War II and on into the 1950s there was a growing restlessness and dissatisfaction with the positivist model of science. Some philosophers tried to amend the model to make it more friendly to existing scientific theories, and a short-lived neoempiricist school in philosophy of science enjoyed a brief celebrity, mostly through the works of Hempel and Quine.

Sometime in the late 1950s, however, a new breeze began to blow across the waters where philosophers of science toiled. This new breeze in philosophy of science swept out the previous concentration on the analysis of mature theories at a relatively abstract level and replaced it with a kind of philosophy of science centered in the history of science and the real workaday experience of actual scientists. The new focus on historical evidence turned up results that were rather unsettling to a positivist or neoempiricist philosopher of science, for those results suggested that the psychological processes of scientific belief formation, change of theories, and the promulgation of scientific knowledge across generational lines are not nearly as rational as the various positivist and neoempiricist schools of thought would have us believe. Actual historical science is much messier—psychologically, sociologically, and politically—than the pure and pristine reconstructions of the positivists would otherwise suggest. The positivist model painted a picture of the workaday practitioner as a "little philosopher," maximally rational, logically astute, immune to irrelevant outside pressures, dedicated to the pursuit of knowledge above all else, and devoid of the foibles and quirks that beset other mortal beings. But a close look at the real life histories of scientists and their careers did not seem to bear out this happy picture of a community of learned minds practicing their craft with granite integrity. Rather, the historical record showed that scientists were often vain, conceited, willing to fudge data, willing to argue unfairly, willing to use larger power sources in society to further their own interests and beliefs, and that, in general, they were *not* little philosophers who are aware of all the epistemological, methodological, and ontological aspects of their chosen fields.

The leading voice of this new historicism in the philosophy of science was the American historian and philosopher of science Thomas Kuhn. Kuhn constructed a powerful general model of science that was a complete alternative to the previous positivist and neoempiricist models of science. Perhaps it was the timing (1962—the beginning of the "Swinging Sixties"), but whatever the cause, the publication of Kuhn's *The Structure of Scientific Revolutions* constituted one of those rarest of events in philosophy: a philosophical work with an even larger impact outside professional academic philosophy than inside professional academic philosophy. Kuhn's general model of systematic inquiry struck a chord throughout the intellectual community, from social scientists to literary critics to poets and writers and physical scientists. There are many professional intellectuals in nonphilosophical disciplines who know virtually no philosophy—except that they know (or think that they know) Kuhn's views about science. In short, Kuhnian philosophy of science became instantly faddish and experienced the fate of every victim of a fad: distortion and misuse. Everybody and their aunt and uncle felt free to cite Kuhn, to claim that Kuhn's views supported whatever their own pet position was, and to argue for their views by presupposing Kuhnian principles, as though the truth of those principles could hardly be doubted by any bright and well-educated person.

One purpose of this chapter is to attempt to rectify the enormous distortion of Kuhn's views that has slowly grown all too familiar during the ensuing four decades since the publication of his masterpiece. When a philosopher's views get used by virtually all sides to a dispute, we can be pretty confident that one or more of the sides have misunderstood that philosopher's views. Kuhn's philosophy of science, I shall argue, cannot be all things to all people. In general, I shall argue that it is not such a radically revolutionary philosophy of science as the popular caricature of it suggests. Instead, I shall argue that it is not even clear that Kuhn's model of science must commit him to broad-scale antirealism. That will be a shocking claim to those avid antirealists who proclaim Kuhn to be one of their most hallowed saints and fellow travelers in the cause.

Kuhn's book, hereafter abbreviated as *SSR*, is not nearly as easy to understand as its popularity might indicate. One helpful way of approaching *SSR* is to see it as composed of two separate projects. First, it contains a brilliant negative critique of the positivist and neoempiricist models of science. Perhaps the case against the positivist model has never been so succinctly formulated as it is in *SSR*. Second, Kuhn presents an alternative model of science that is meant to replace the destroyed positivist model. Most nonphilosophers are more impressed with the second project than they are with the first. It is Kuhn's replacement doctrine—what he says science is, rather than what he says it is not—that strikes a vital nerve in them. On the other hand, most philosophers are more impressed with the first of Kuhn's projects in *SSR* than they are with the second. That is because philosophers have a well-honed appreciation for why things do *not* work (such knowledge is far from unimportant to them—it may be more important than knowing why things do work), and also because philosophers understand the highly questionable downstream consequences of Kuhn's replacement model better than do most nonphilosophers. What makes *SSR* a complicated read is the fact that Kuhn skillfully intertwines the two projects, showing the reader how the failures of the positivist model lead naturally (or so Kuhn claims) to his own replacement view of science. We will begin our exploration of *SSR* chronologically. We will start where Kuhn starts in the text: with a critique of the **incrementalism** to which the positivist model is committed.

7.1 Anti-Incrementalism

The positivist model of science and its offspring are committed to an epistemological position that Kuhn labels incrementalism. Incrementalism holds that, in science, over time, there is a slow but continuous accumulation of an ever-increasing stock of truths. Roughly, later science gets more correct than earlier science; and, what science gets correct once, stays correct once and for all—science doesn't "go backwards," it doesn't lose truths once they have been discovered.

Incrementalism strikes most people as obvious. Surely immunologists today in 1996 know much more about the real nature of allergic phenomena than the community of immunologists knew in 1970; and, those 1970 immunologists knew more about allergic phenomena than the immunologists of 1940, and so on. Furthermore, it would be rather bizarre if, short of some apocalyptic world crisis that set all culture back a thousand years, important pieces of knowledge about allergic phenomena were to be lost, to vanish, to cease to be used and depended on by practicing immunologists. Could it really happen that, for example, the truth that tissue mast cells contain granules of histamine was lost from immunological science? What would that be like? How could it happen? (If the example is too specialized for the reader, then we may use a more mundane case: Try to imagine chemical science losing the truth that the chemical composition of water is H_2O.) Nevertheless, Kuhn insists that, first, a brutally honest look at the history of science will reveal a number of cases in which the evolution of knowledge within a scientific domain of inquiry has been discontinuous—where the development of knowledge within a scientific field backtracked, and knowledge was temporarily lost that had been previously gained—and, second, that understanding how this can happen involves adopting an orientation to philosophy of science completely different from that of the positivist model. This new orientation centers philosophical investigation less on the rational reconstruction of mature scientific theories—after the fact, as it were—and more on the idiosyncratic history of development within a given scientific domain. Philosophy of science ceases to be so much applied logical analysis and becomes more like history of science. That history, Kuhn claims, shows that it is unhelpful to analyze science using the concept of a theory as the core unit of analysis. Theories, in the sense that the positivist model understood them, come late in the development of a scientific field; and, in an important sense, the daily practice of a given scientific discipline is not especially bound up with the theories associated with that field. Practitioners do not carry out their daily activities with theories as much as they carry them out with a different organizing concept, a concept justly famous (or infamous, depending on your point of view) for both its richness and its vagueness. With this new organizing concept, Kuhn was able to construct a profoundly appealing model of how a given scientific domain of inquiry develops from its infancy through several distinct stages to full maturity, and even beyond that as it periodically experiences upheavals he calls "revolutions." A scientific field has something like a life history—an infancy, an adolescence, and a series of different adulthoods (brought about by occasional revolutions). The key to understanding this life history is Kuhn's new organizing concept.

7.2 Paradigms and the Practice of Normal Science

Kuhn argues that a mature scientific domain is organized around something called a **paradigm**. The word 'paradigm' is Greek for 'pattern'. This term denotes the central object of study in Kuhn's philosophy of science, for all the action in the history of science is related to the rise and fall of paradigms. Saying just what a paradigm is, on the other hand, is notoriously complicated. The philosopher Margaret Masterman, soon after the publication of *SSR*, wrote a review of the book in which she argued that the term 'paradigm' was used by Kuhn with 22 distinct meanings, which could be grouped into three larger families composed of interrelated concepts. The nasty interpretation of Masterman's review was that she exposed the fact that Kuhn had used a core term which was 22 ways ambiguous. The kinder interpretation was that he'd only used a core term with a three-fold ambiguity. This was no minor pedantic charge by Masterman. The more ambiguous and vague a concept is, the less definite is the content it has and therefore the less we say by using it. A concept so broad as to include everything under itself would be literally meaningless: It would have no content precisely because it ruled nothing out as falling under it.

Kuhn certainly does spread out his remarks about paradigms so that the concept covers a lot of ground. It is not clear which remarks about paradigms indicate their more essential features and which indicate their more accidental features. According to Kuhn, a paradigm,

(1) is centered around an overtly recognized achievement that generates model problems and solutions for a community of practitioners
(2) creates a tradition of experimental procedures, techniques, and instruments that licenses what problems ought to be worked on and what would count as solutions to them
(3) renders possible a set of repeatable and standardized illustrations of various theories as revealed in textbooks, lectures, and other artifacts of professional training
(4) establishes what qualifies as acceptable examples of successful scientific practice
(5) brings into existence a network of theoretical, instrumental, and methodological commitments
(6) constitutes a metaphysical worldview
(7) cannot be characterized by precise rules or sentences (hence, it is not formalizable in the spirit of positivism)
(8) is not the same as a theory, although a paradigm may include theories.

Do these characterizations help us to pin down the notion of a paradigm in science? If not, perhaps some examples of paradigms would help. Kuhn treats the following as examples of paradigms in the intended sense:

(i) Newtonian mechanics
(ii) Einsteinian mechanics (relativistic)
(iii) Daltonian chemistry
(iv) the fluid flow theory of electricity
(v) Copernican astronomy.

A simple question naturally arises at this point. Why not treat these as theories, like everybody else does? Kuhn's answer is, because (i) to (v) are much more than theories; for each of them, in addition to including theories, constitutes a way of practicing research. Each represents a set of unique attitudes toward data, and each of them brings a distinct worldview with it. Each of them is distinguished from its competitor paradigms in the same field by suggesting a slightly different set of problems to be worked on and what would count as solutions to those problems. Each of them generates a distinct pedagogical tradition, a way of educating succeeding generations of practitioners, a particular way of writing textbooks, engaging in communication within the community of practitioners, and so on.

A general approach to research comes to dominate a field—becomes a paradigm—when the practitioners working under its direction score an amazing research achievement, an achievement everyone recognizes as such, even those practitioners committed to competing approaches (although they often do so grudgingly). Prior to any one research orientation scoring such an achievement, Kuhn says that the field of practice in question is in its "preparadigmatic stage." What counts as scientific practice in the preparadigmatic stage of a science is very different from research done under the direction of a paradigm. Kuhn calls the former "natural history." What he means is that, minus an organizing paradigm that can define practice by presenting an achievement to be imitated, a field of science consists of unstructured, random fact gathering. Those working in a preparadigmatic field are not even sure what count as important data and what data are irrelevant. They don't agree on what needs explanation and on what can be taken for granted in that field. The result is an inability to filter out data so that only the explanatorily relevant facts are investigated. Hence, the preparadigmatic field is dominated by the heaping up of large numbers of seemingly disconnected "facts"—seemingly disconnected because there is no paradigm to organize the relevant facts among them into a theoretically useful hierarchy, and to throw out the irrelevant ones in the process. One way in which Kuhn characterizes this lack of structure in a preparadigmatic field is to say that one of the things settled by the arrival of a paradigm is what counts as the fundamental principles of that field. *Paradigms fix the fundamental principles of a domain, so that they need not be questioned further. Once a paradigm has arrived, research need not be directed to settling the fundamental principles—they have already been settled.* Until a paradigm arrives on the scene, therefore, the practitioners in a scientific domain argue over fundamental issues. The most basic principles of the domain are open game. We might describe this state of affairs by saying that the field is divided up into "schools" of competing "enthusiasts," scientific amateurs who are struggling to give birth to real science, to turn their unsystematized sleuthing into precise experimental research. Under preparadigmatic conditions, little *precise* research can be carried out, for precise research depends on being able to take fundamental principles for granted and focusing on the minute, detailed "puzzles" of whatever systems are under study in that domain.

The triumph of a paradigm in a domain that was previously without one produces a new form of scientific practice that Kuhn calls **normal science**. Normal science is very different from preparadigmatic science, and that difference is due to the presence of a dominating paradigm. The paradigm generates a sense of security for the practitioners through having settled the domain's fundamental principles, and by promising to each practitioner a career's worth of interesting puzzles to work on, puzzles that have definite solutions. Hence, normal science quickly seeks to institutionalize itself through the creation of professional journals only accessible to practitioners, professional associations only accessible to practitioners, and a pedagogical tradition through which the paradigm can be taught to succeeding generations. The settling of fundamental principles also brings in its wake something like a shared worldview, which allows practitioners to obtain a set of standard responses to any questions regarding the metaphysical basis of their domain. I believe we now have enough Kuhn digested to be able to give a rough characterization of the concept of a paradigm. For our purposes, let us understand the term in the following sense.

> A paradigm is an achievement that defines practice for a community of researchers. It defines practice because the achievement constitutes a model to be imitated and further extended—future research tries to fit itself to the same pattern as the original achievement. This definition of practice brings in its wake: the settling of the fundamental principles of the domain, the subsequent possibility of extremely precise research, a pedagogical tradition that trains succeeding generations in the use of the paradigm, a collection of institutions designed to promote the paradigm (professional journals, professional associations), and a worldview with metaphysical consequences.

A paradigm ushers in a period of normal science in the domain in question. To the internal structure of normal science we turn our attention in the next section.

7.3 Puzzle Solving and the Cautiousness of Normal Science

Normal science is usually not very exciting by Hollywood standards. No one makes movies about the practice of normal science, and very few novels have been written about it. Normal science is inherently cautious, irritatingly slow, and obsessively dedicated to solving puzzles—the unearthing of "little" details left over for ordinary research by the initial achievement of the paradigm. The paradigm settled all the *big* issues, all the main issues. What it did not do was fill in all the blanks, the nooks and crannies. There is much "mopping up" work to be done. It is this mopping up work that constitutes the daily grind of normal science. Mop-up work is intrinsically highly detailed. The paradigm suggests that certain aspects of the phenomena studied in the given domain will be revealed only after the practitioner has delved into them in the required detail. Hence, much of the time normal science consists of research that strikes the outsider as obnoxiously specialized and saturated with an impenetrable jargon that is virtually incomprehensible in terms of its esoteric technicality. The purpose of normal science is to ferret out the "hidden" details at the microstructural level of the phenomenon under study, and that will require that the practitioner focus in on a minute piece of the overall phenomenon with a level of intensity and magnification that seems almost perversely obsessive to the nonpractitioner. Frankly, it is this author's opinion that the average nonscientist (and I include most well-educated humanists in that group) has no idea of how positively tortuous the causal story often turns out to be in the microstructural details. Doing normal science takes an amount of patience with, and tolerance for, overwhelmingly detailed complexity that the average nonscientist simply does not possess in the requisite amounts.

Kuhn gives a classification of different kinds of normal science research that some philosophers of science, including this author, have not found helpful (to wit: "refining" facts the paradigm suggests are important, matching facts to predictions, and "extending and articulating" the paradigm). Of more interest is Kuhn's claim that a typical normal science research project is *not* intended to challenge the reigning paradigm under whose direction it is carried out. Normal science does not seek to overthrow the reigning paradigm. It is rather a time of peace and contentment, theoretically speaking, for those practitioners engaged in inquiry under the direction of the paradigm. There are almost no practitioners who are especially unhappy or dissatisfied with the reigning paradigm. There is virtually no fomenting discontent, virtually no desire to stage a revolution and find some new general orientation within the domain in question. This security and peace allows the practitioner the luxury of "taking a chance" and digging deeply into the microstructural underbelly of the domain. The practitioner can be pretty confident that five or six years spent ferreting out some minor piece of microstructural detail won't be washed away to nothing by a sweeping revolution and overthrow of the paradigm that alters all the fundamentals in the field, including the fundamental principles presupposed by those five or six years of research. Hence, normal science encourages a practitioner, for example, to build an elaborate and expensive machine that will measure some important theoretical constant of use in the domain. A physicist might spend three years building and using such a machine to measure the speed of light to five more significant figures to the right of the decimal point. Or, closer to our chosen field of illustration, an immunologist might spend two years creating and using an elaborate method to sequence the amino acids that compose the protein chain of the CD56 surface marker on the surfaces of NK cells. Only because the physicist and the immunologist can take the fundamentals of their fields as settled by their respective reigning paradigms, only because they don't have to build their fields from the ground up with each experiment, can they be allowed the time and expense to engage in such unveiling of microstructural minutiae. A glance through a professional-level research journal is instructive on this point. The inaccessibility of the research articles in such a journal to all but other specialists is usually fairly obvious. Readers skeptical of the inaccessibility are referred to Figures 7.1 and 7.2. Knowledge of deeply technical minutiae is what constitutes professional-level expertise in the field in question.

Paradigms do something else besides generate minute puzzles to be worked on—they put constraints on what will count as successful solutions to those puzzles. They promise not only a puzzle to solve, but a definite way of telling when one has solved it successfully from when one has not solved it successfully. Recall that the paradigm was an achievement of some kind, a stunningly successful solution to some interconnected set of empirical problems. That solution becomes the model solution for subsequent practitioners under the paradigm to emulate. Not every surmised solution to a puzzle is the right solution. Not any

The Daily Blurb

Scientists Discover!

"Mercury-Based Cuprate High-Transition Temperature Grain-Boundary Junctions and SQUIDS Above 110 Kelvin"

The Daily Blurb

Scientists Discover!

"Site-Specific Cleavage of Human Chromosome 4 by Triple-Helix Formation"

The Daily Blurb

Scientists Discover!

"Deregulation of a Homeobox Gene, HOX11, by the t(10;14) in T Cell Leukemia"

The Daily Blurb

Scientists Discover!

"Identification of the Envelope V3 Loop as the Primary Determinant of Cell Tropism in HIV-1"

The Daily Blurb

Scientists Discover!

"Raman studies of Alkali-Metal Doped A_xC_{60} Films (A=Na,K,Rb, and Cs; x = 0,3, and 6)"

The Daily Blurb

Scientists Discover!

"Formation of a Monomeric DNA Binding Domain by Skn-1 bZIP and Homeodomain Elements"

The Daily Blurb

Scientists Discover!

"Arrest of Motor Neuron Disease in *Wobbler* Mice Cotreated with CNTF and BNDF"

The Daily Blurb

Scientists Discover!

"Interleukin-2 Receptor γ Chain: A Functional Component of the Interleukin-7 Receptor"

The Daily Blurb

Scientists Discover!

"Delocalization of Vg1 mRNA from the Vegetal Cortex of *Xenopus* Oocytes After Destruction of Xlsirt RNA"

The Daily Blurb

Scientists Discover!

"PHAS-1 as a Link Between Mitogen-Activated Protein Kinase and Translation Initiation"

Figure 7.1. The minute detail of normal science: titles of sample research papers from the journal *Science*.

```
  1 CTAAGAAAGGGATCCGGGAGAATGGCCATAGGACACTTCCAA
                        ↑me al  il  gy  hi  ph gn     7
 43 AGTCTAAATCGAAAGGCTTTTGATGGGATCCCAGCCTCCTGC
    se  le  an  ar  ly  al  ph ap gy  il  pr  al  se  cy   21
 85 GTCAACGGTCAAGATATGTGCCGAAAACTTTACTTGACCGCC
    va  an  gy  gn  ap  me  cy ar  ly  le  ty  le  th  al   35
127 TGGGAACTTTACGAAGTTATATCACCACTTAGAGAGCTTACC
    tr  gl  le  ty  gy  va  il  se  pr  le  ar  gl  le  th   49
169 GGTAGTAAATCCATGTGGGGGACTTTGGAAATACCCTGCTCT
    gy  se  ly  se  me  tr  gy  th  le  gl  il  pr  cy  se   63
211 TTTGCCTCACATACCCCAATAGCTAATAAACCGGAGATTGAT
    ph  al  se  hi  th  pr  il  al  an  ly  pr  gl  il  ap   77
253 CATAATATCCGACCTCCAAGGGAGAGTGGTTTTTCCCATTGT
    hi  an  il  ar  pr  pr  ar  gl  se  gy  ph  se  hi  cy   91
295 CTTCAAACCAACACTCGCTGGTCTCCGCAGACATACAGATAC
    le  gn  th  an  th  ar  tr  se  pr  gn  th  ty  ar  ty  105
337 AAACAGTGTCATGAAGTCATTAGTTTTAACGATGCCGTTATG
    ly  gl  cy  hi  gl  va  il  se  ph  an  ap  al  va  me  119
379 CAATACAATGCCTTCATCATGTGGGATTGCCCCTCTCTTTGG
    gn  ty  an  al  ph  il  me  tr  ap  cy  pr  se  le  tr  133
421 TGCCATGAGATGAATACCAGAGATCAGTAACCGAGGCAG...
    cy  hi  gl  me  an  th  ar  ap  hi  *  ↑               142
```

This fictitious gene codes for a protein about the size of a small cell-surface receptor. The arrows indicate the initiation and termination codons which start and stop the transcription of the gene. The four DNA bases are A = adenine, C = cystosine, G = guanine, T = thymine. The genetic code is read in triplets called codons, every three-base sequence codes for an amino acid except for the three termination codons (the initiation codon is also the only codon for the amino acid methionine). The number of bases is indicated down the left side, the number of amino acids in the protein is indicated down the right side. This gene contains 429 transcribed bases coding for 142 amino acids plus the termination signal *. The dots indicate further bases untranscribed due to the preceding termination signal. The amino acid abbreviations are: al = alanine, an = asparagine, ar = arginine, ap = aspartic acid, cy = cysteine, gl = glutamic acid, gn = glutamine, gy = glycine, hi = histidine, il = isoleucine, le = leucine, ly = lysine, me = methionine, ph = phenylalanine, pr = proline, se = serine, th = threonine, tr = typtophan, ty = tyrosine, va = valine.

Figure 7.2. A standard presentation of a gene such as might appear in the sorts of papers mentioned in figure 7.1.

old result of a research experiment is the result that the paradigm says is to be expected. But nature is sloppy, complicated, and structurally obstinate, and our powers of inventiveness and precision are finite and fallible. The outcome of this is that no paradigm is perfect, no paradigm can give a complete account of the domain over which it reigns. The intrinsic incompleteness of paradigms means that there will always be a few research results that do *not* fit the paradigm—that is, research results that conflict with what practitioners using the paradigm would expect to observe. These research results which don't fit under the reigning paradigm Kuhn calls **anomalies** (from the Greek, 'anomos', meaning 'not lawful'). *Normal science inevitably produces anomalies.* This inevitability is not hard to understand given a moment's reflection on the matter. It is a function of the level of greater detail at which normal science operates. As microstructural complexity increases, so does the probability of finding something that doesn't mesh with the reigning paradigm. The paradigm, after all, was largely silent about the microstructural details below a certain level of complexity—for the paradigm paints in broader, larger-scale strokes (it has to, if it's going to settle fundamentals, provide a general worldview, be an imitable achievement, and so forth), and it cannot anticipate all the details of microstructural reality to be found in the phenomena of the domain in question. In a way, the very strength of normal science is also its greatest weakness: *The investigation of phenomena in excruciatingly fine-grained detail is the primary source of the anomalous data which eventually destabilize the reigning paradigm.* The end of the last sentence should clue the reader in to the reason why anomalies play a vital role in Kuhn's model of science. To the investigation of anomalies we now turn.

7.4 Anomalies

Anything built by human beings is an artificial device, and paradigms in science are no exception to this general principle. Artificial devices fit into the natural world more or less smoothly, but never perfectly. No matter how powerful and seemingly invincible a paradigm is, the production of research anomalies under its rule is inevitable. What happens to these anomalies, these results that don't mesh with what the paradigm says practitioners should have observed? The answer Kuhn gives to that question is contentious and bound to alienate those who hold an exalted view of science: The vast majority of anomalies during times of normal science get ignored—well, perhaps that is a bit strong. They are usually acknowledged as anomalies by practitioners, but the majority of practitioners do not take the occurrence of anomalies within their domain of inquiry as indicating that something is seriously, fundamentally wrong with the paradigm. During times of normal science, anomalies are not treated as cases that falsify the paradigm, the way a strict Popperian falsificationist presumably would treat them. Anomalies are noted, filed away in the professional journals and research reports, and for the most part forgotten for the time being. There is simply too much other normal science research to be done—too much else is *non*anomalous for practitioners to throw in the towel and revamp their entire field from the ground up. The psychology of practice is complex at this point. Practitioners are not exactly being dishonest—they are willing to recognize that the anomalies are anomalous—but rather, there is a kind of research inertia at work. To see Kuhn's point, the reader should imagine that he or she has invested eight or more years and perhaps hundreds of thousands of dollars in professionalized scientific training, training in a particular paradigm within the reader's chosen field of study.

The reader has accordingly committed his or her time, money, effort, and study to approaching the phenomena of interest in the domain using a specific paradigmatic orientation. The reader had perhaps published research articles and papers that presuppose the correctness of the paradigm in question, attended conferences and meetings where the proceedings presupposed the correctness of the paradigm in question, and secured academic employment to teach the field as understood under the paradigm in question—in short, the reader's entire professional identity, standing, reputation, and accomplishments, all depend on the correctness of the paradigm within which the reader was originally trained. Now, along comes some young whippersnapper, some reckless troublemaker, some attention-seeking upstart in the reader's field who announces an anomalous experimental result. What will the reader's reaction be? Well, the last thing the reader would do is throw up his or her hands and declare his or her whole career's worth of work suspect. What Kuhn is pointing out is that there are institutional reasons why normal science is inherently "conservative" in the face of anomalies. No practitioner who has built a career out of work that depends on the correctness of a particular paradigm wants to see that career's worth of work questioned under some different paradigm that might very well render the research worthless, or at best, require that the work be repeated under slightly different conditions of evaluation. No wonder anomalies get ignored during normal science. Anomalies kind of pile up during normal science, not really bothering anybody very much, but they are still there, recorded, and waiting for the moment when they will gain center stage and begin to torment the majority of practitioners.

Kuhn is concerned to note a second reason why anomalies get ignored: Allegiance to the reigning paradigm cognitively blinds practitioners from seeing the obvious way to resolve their anomalousness. Recall that a paradigm has a holistic aspect to it—it affects everything the practitioner does in the course of research. It even alters the way the practitioner "sees" the observational data. The practitioner sees the data a certain way because prior training in the paradigm has "cut off" from cognitive access alternative orientations under which something slightly different would be observed. Kuhn presents a famously rich discussion of this feature of normal science, using the historical case of the discovery of oxygen. Kuhn's argument that Antoine Lavoisier was able to "see" oxygen as the distinct chemical species it is, whereas Joseph Priestley, committed to a different paradigm in chemistry than Lavoisier, was not able to "see" oxygen as anything but "dephlogisticated air" had an enormous impact on subsequent discussions of the observational/theoretical distinction. The argument makes Kuhn's point that different research results are anomalous within different paradigms. What counts as an intractable anomaly under one paradigm might be the very core of what a different paradigm best explains. It is in this way that a paradigm affects the practitioner's worldview: What needs explaining, what is mysterious, and what is commonplace can fluctuate with paradigm change.

Do anomalies get ignored forever? Not usually. The history of science suggests that eventually anomalies pile up until they reach a kind of critical mass, to borrow a metaphor from nuclear physics. Eventually, a generation of practitioners arises, which, for its own idiosyncratic reasons and motives, begins to take the accumulated store of anomalies seriously. Kuhn claims that no hard and fast generalizations can be made about when this happens. No particular number of anomalies must accumulate. No particular amount of time must pass. No particular kind of generation of practitioners, psychologically speaking, must arise. This rebellious generation is bothered by the anomalies. It interprets them as indicating that something is basically incorrect about the reigning paradigm. The first change this produces in normal science is a change in the orientation of research. Normal science research is usually not directed at challenging the basics of the reigning paradigm. As a field of science goes into crisis preparatory to experiencing a revolution, more and more research begins to be deliberately aimed at challenging the fundamental principles supposedly settled by the reigning paradigm. The reigning paradigm in some cases comes under direct research attack, and experiments are designed with the express purpose of falsifying the paradigm outright. Accordingly, two interesting features characteristic of what we might call "crisis science" appear: Fundamental principles cease having their formerly secure status; and practitioners begin to tinker with the paradigm, constructing alternative and incompatible versions of it to accommodate some of the anomalies. Kuhn is especially insistent that the sudden appearance of many alternative and incompatible versions of a paradigm that previously reigned in monolithic majesty is a sure sign that a field of science has gone into crisis and that the comforts of normal science have ended for practitioners in that field. Sometimes the reigning paradigm is sufficiently rich in conceptual power and sufficiently flexible in application that it can weather the crisis without falling from dominance. In such cases the reigning paradigm pulls through, not without some scars, and probably not without some changes to its former version of itself. But, every once in a while, the reigning paradigm does not succeed in staving off defeat. The crisis is too much for it and it falls from dominance. When that happens a scientific domain has passed from normal science, through crisis, all the way to a period of what Kuhn calls **revolutionary science**. Revolutionary science is the kind of science that they do make movies about, at least occasionally, and that is because it is the one kind of science in which high philosophical drama is at home.

7.5 Revolutionary Science

When a domain of inquiry undergoes the experience of revolutionary science its fundamental principles are up for grabs. The research atmosphere in the domain becomes rather heady and uncertain. Excitement and turmoil ripple throughout its community of practitioners. Different practitioners take different sides on the same disputed theoretical issues. They split over allegiance to different suggested reorientations of the whole domain. Soon, some brilliant practitioners suggest a replacement paradigm, a way of restructuring everyday practice in the domain so as to constitute normal science once again—a way of returning to puzzle-solving research but perhaps with new puzzles, new solutions, and new methods. But the old paradigm still has its defenders. The "Old Guard" does not die easily. Paradigms do not go gently into those good nights. Usually, the revolutionaries agitating for the new paradigm are the younger practitioners—those with less to lose from a complete overhaul of the domain—while the defenders of the old paradigm are the older practitioners, those who have put in nearly a whole career's worth of work under the threatened paradigm. The defenders are generally stubborn, impervious to argument, and adamant in their refusal to bury the old paradigm. Kuhn famously depicts the irrational elements at play in this situation, how the Old Guard stoops to trying to thwart the careers of the younger revolutionaries. The Old Guard tends to control access to research money, to publication space in the professional journals of the field, and to academic appointments at institutions of higher learning. It can get nasty, petty, and downright unpleasant. Scientists have enormous egos, and nothing less than a lifetime's work can be at stake in a revolutionary crisis. What I have been describing for the last several sentences is really a kind of "war" over which of the competing paradigms will take command of the domain. Kuhn himself likens the professional fighting that takes place during a **scientific revolution** to a political dispute. The parties resort to unfair and ugly means to achieve victory—stifling brilliant research by the other side, gutting academic careers, refusing to grant research funding for worthy projects. Rarely does reasoned argument play a role in any of this. The parties to the paradigm war act more like ruthless local politicians pursuing their own self-interests than like philosophers dedicated in Socratic fashion to the pursuit of truth no matter what the personal cost. Kuhn claims, in one of his more extreme moments of inspiration, that the experience of switching to a new scientific paradigm is similar to a "quasi-religious conversion" experience, somewhat like a sudden Gestalt experience, in which the practitioner comprehends the appropriateness and superiority of the new paradigm all at once, "in a flash," without being able to decompose the experience into discrete and rationally defendable parts.

Kuhn has a doctrinaire reason for holding that paradigm conversion is nonrational: He claims that competing paradigms are **incommensurable**. That is a long word that means basically what 'noncomparable' means. The term has a well-defined mathematical meaning: Two quantities are incommensurable if they cannot be measured using a common standard of measurement. Different paradigms, claims Kuhn, cannot be directly compared with each other; for direct comparison would require that there be some third "superparadigm" from whose standpoint we can look down and compare the first two paradigms with a common scale of assessment. But there are no such superparadigms, and assessment itself is paradigm relative. The practitioner is always operating from the perspective of some particular paradigm. The practitioner can never get outside all paradigms, can never reach some paradigmless state of conceptual neutrality, to be able to compare fairly and without bias the respective merits of the warring paradigms between which a choice must be made. You have to use a paradigm to evaluate a paradigm, and that makes all such evaluations biased in favor of the paradigm that is used to make the evaluation.

Now if paradigms are incommensurable, then paradigm conversion is not likely to be a rational process; for reasoning itself is a paradigm-relative activity—or so claims the Kuhnian—and what counts as a good reason for adopting a certain paradigm will differ in different paradigms. Hence, all that is left is the quasi-religious conversion experience. Allegiance to a paradigm in the midst of a revolutionary period of science is more a question of "politics" than of evidential reasoning, more a matter of who is in power, controls the research money, writes the textbooks, makes the academic appointments, edits the professional journals, and teaches the young. Epistemologically irrelevant factors such as nationality, language spoken, desire for fame and admiration, and monetary gain can at times play pivotal roles in determining the outcome of a scientific revolution. Or, so Kuhn seems to suggest in the more loosely written sections of *SSR*. It has always been a matter of some contention just how seriously Kuhn meant for his readers to take this deflationary portrait of the scientific practitioner under the pressures of revolutionary crisis. Kuhn is convinced in certain passages in *SSR* then when two practitioners adopt widely differing paradigms in the same field, they practice their crafts in "different worlds"; for a paradigm is what organizes disconnected experiences into a coherent world of orderly practice. The incommensurability principle can then be understood as locking the two practitioners into their separate worlds. They can't establish a common ground of communication, and Kuhn claims that often in the history of science it has happened that the warring parties in a paradigm war "talk past each other." The two sides may appear to use the same specific technical terms when they attempt to settle their differences through open debate with each other, but they don't both mean the same thing by the use of those shared technical terms; and so, the attempt at rapprochement fails through a kind of blind, group-equivocation fallacy. To use a quaint cliche, it is not merely that one side says to-may-toe and the other side says to-mah-toe; rather, one side refers to the tart red fruit used in pizza by the use of the term 'tomato' and the other side refers to some distinct kind of food stuff by the use of the term 'tomato'—a food stuff that is perhaps not a fruit but some other kind of food subtly different from ordinary tomatoes. But the warring sides often fail to appreciate this difference. They all believe themselves to be using a common vocabulary in the course of their disputations with each other. Such a linguistic impasse dictates that conversion must be largely nonrational, not brought about through rationally articulated discourse and argument, but through a kind of highly subjective Gestalt "enlightenment" experience, different for every convert, and not reconstructible after the fact in a way that can render it completely rational and evidence based.

Once enough such conversions have taken place in favor of one of the competing paradigms, if that paradigm's devotees are skillful enough, they can succeed in basically shutting down research under all the other competing paradigms. The devotees move into positions of institutionalized power. *They* get to write the next generation of textbooks, to edit the professional journals, to hand out the research money, and to make the academic appointments. The end result is a cessation of the revolutionary crisis and a return to another period of normal science, this period being dominated by the reign of the newly triumphant paradigm. Hence, over large spans of time, a given domain of scientific inquiry undergoes a cyclical evolution in which long periods of normal science are interrupted by a number of shorter periods of revolutionary science. At this point Kuhn brings into the picture the anti-incrementalism with which he begins *SSR*. In general, Kuhn claims *there is no guarantee that later periods of normal science preserve the "truths" of earlier periods of normal science.* Later periods of normal science may be inferior in certain ways to earlier periods (although, Kuhn acknowledges that in most cases later normal science is an improvement on earlier normal science). There is no guarantee that truth slowly accumulates over time, that later paradigms are necessarily more accurate representations of real phenomena than earlier paradigms. We can represent this thesis of Kuhn's by the use of a helpful graph. Figure 7.3 illustrates the thesis that normal science is distinguished from revolutionary science by, among other things, scarcity of research that directly challenges the reigning paradigm. Plotting on the vertical axis the percentage of research devoted to direct challenges to the reigning paradigm against time on the horizontal axis, the progress of science in a given domain is cap-

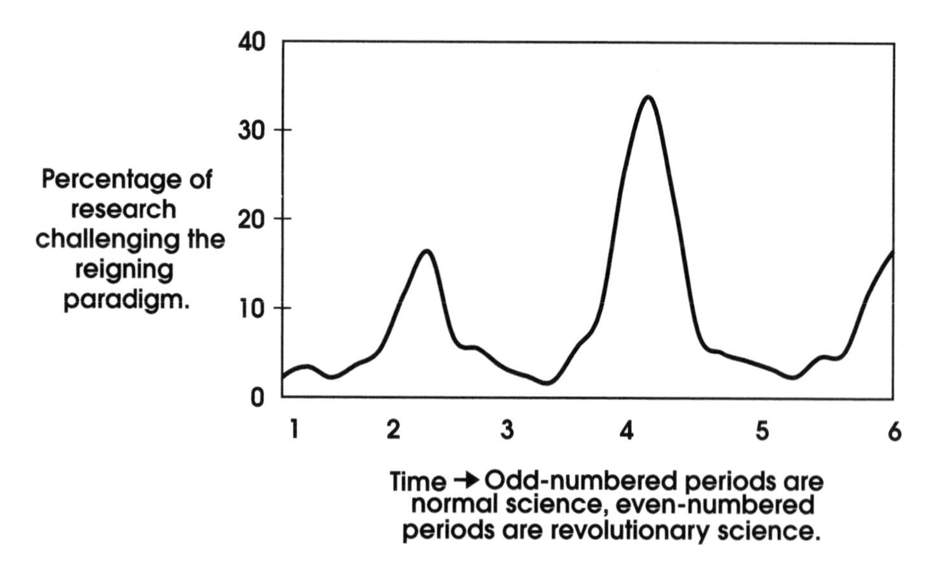

Figure 7.3. Differences in the nature of research over time within a scientific field under the Kuhnian model of science.

tured by the graph. Notice that there is no uniformity to the length of revolutionary stages or to the length of normal science stages. Kuhn is quite clear on this point. The evolution of science obeys no temporal regularities. There are no hard rules about, for example, how many anomalies must accumulate before a generation of revolution-minded practitioners arises. Nor are there rules about how many conversions are required before a new paradigm gains dominance. It is a fuzzy and sloppy business, the evolution of science, and only after the fact, as history of science, can it be pieced together into a coherent story. In Figure 7.4 I present a flow-chart diagram of the Kuhnian model of science that I hope will be helpful to the reader (with thanks to Dick Leatherman for the idea of using a flow chart).

Kuhn makes clear that there is a price to be paid for the benefits of normal science. First, each period of normal science ends up rewriting the history of its domain from its own perspective. Kuhn argues that the average practitioner of a science is generally quite ignorant of the *real* history of his or her own field. That is because each paradigm-bound generation of textbook writers distorts the past of the field so as to make it appear that previous generations of practitioners were really working on the same problems, with the same methods, as the present generation. Each paradigm, in other words, rewrites the history of the field to make itself out to be the natural evolutionary result of what has gone

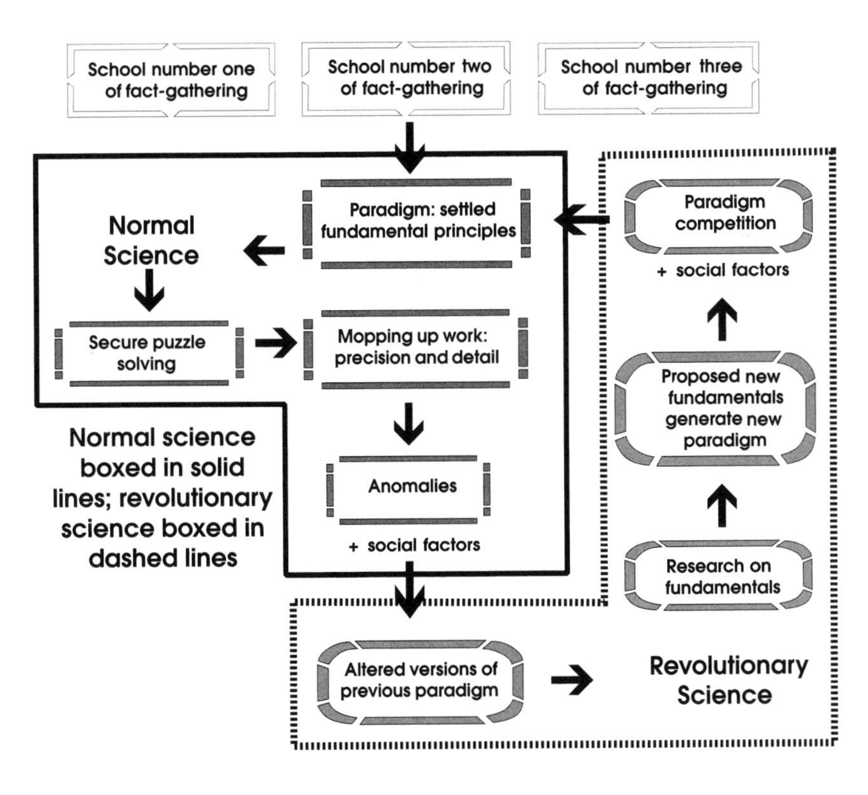

Figure 7.4. A flow-chart representation of the Kuhnian model of science.

before. But this generally distorts the real history, Kuhn insists, and *SSR* is justly famous for containing a number of interesting cases illustrating this sort of distortion of history (for example, a number of chemistry texts attribute the introduction into chemistry of the concept of a chemical element to Robert Boyle, but in actual fact Boyle only mentioned chemical elements in the course of claiming that there are no such things). Second, normal science, by having settled fundamental principles and thereby freeing the practitioner to ferret out minute detail, dooms itself to an ever-increasing alienation from the general public. As the scientific details unearthed by precise normal science research become more complicated, esoteric, and laden with jargon, more and more of science ceases to be comprehensible to the general populace. The Person of Science ceases to be a traditional Renaissance Person able to reach out with understanding to all segments of society and becomes more and more of a jargon-spouting expert in technical minutiae who is comprehensible only to fellow experts in the same field. Hence, science passes away from being something absorbable into the general culture of a society and ends up instead a socially encapsulated institution insulated from the hearts and minds of the masses.

But there is an upside to normal science, too. Only normal science could possibly achieve the level of detailed knowledge we now possess concerning the more intricate seams, nooks, and crannies of the natural world. Permanent revolution would not be capable of generating that result, for all the energy of practitioners would in that case be spent arguing over fundamental principles and the details would go forever unearthed. But we ought to ask an important question about this whole story so far. These details I keep speaking of, they would seem to change with every change of paradigm. What counts as "the facts" will be one more paradigm-relative feature of science under the Kuhnian model, will it not? And, if so, what becomes of the traditional realist notion that science discovers the detailed facts, that the detailed facts are what they are independently of any humanly contrived paradigms or theories about them? The Kuhnian model is often accused of having replaced this traditional picture of science as something that *discovers* facts with a picture in which science helps construct the facts; that is, the facts are partially artificial, partially constructed by our paradigm-relative methods of inquiry and practice. On such a picture of science, the natural world is, as the fashionable saying goes, *socially constructed*, at least to some significant degree, if not completely (and the latter possibility strikes many thinkers as out-and-out idealism). This kind of Kuhnian social constructivism is a form of relativist antirealism and, to this serious charge we turn next.

7.6 Antirealism, Relativism, and Their Shadows

The traditional view held by the scientific realist is that the different theories or paradigms of science investigate a common natural universe. The natural world is what it is independently of our particular paradigm-based manner of investigating what it is at any given moment. This otherwise commonsense view of science is threatened by the Kuhnian model of how science operates. That model maintains that there is no method of inquiry that stands outside any and all paradigms. Scientific practice always takes place under a paradigm, it is always filtered through a paradigm, and the paradigm in question infects everything it touches, forcing those who practice under it to see, to understand, the part of the universe under investigation a certain way rather than any other way. Hence, there is no nonparadigmatic form of inquiry in science (recall that the preparadigmatic stage of a field is mere natural history, not genuine science), and only such a form of inquiry—allegedly carried on outside all paradigms—could claim to have discovered a natural world undistorted in its structure by the biases of any paradigm. All scientific knowledge is relative to a paradigm under which it qualifies as knowledge. That is the relativism that seems to follow from Kuhn's picture of science. But relativism naturally begets antirealism by a simple argument from the history of science. First, we note the undeniable historical fact that, in different eras, a given field of mature science has been dominated by very different paradigms. Second, we note that these differing paradigms are not only seemingly mutually inconsistent; but, if Kuhn is correct, they cannot even be compared to the point of saying that they are mutually inconsistent, for mutually inconsistent claims must be claims about a common subject matter, and the incommensurability of paradigms cuts off the ability to ascertain whether the paradigms are about a common subject matter. So, the situation is actually worse than being one of mutual inconsistency among different paradigms; it is a situation we might call *mutual noncorrespondence.* Different paradigms construct different natural universes, and no two of those universes can be said to be similar to any degree, for there is no common measure by which such similarity could be measured. Different paradigms do not correspond to each other in any well-defined sense—at least, not under the Kuhnian model of science.

So, the main question arises: Which of the noncorresponding universes constructed by which of the incommensurable paradigms is the most accurate representation of reality? The devout social constructivist will object to the question itself. What makes you think there is anything called "reality" that exists independently of how our canons of inquiry systematize our experience, the social constructivist will ask? But that position is simply antirealism. Kuhn himself was notoriously hard to pin down on this issue in *SSR*. Somehow he managed to skate around it without ever coming directly to terms with it. It is standard to interpret Kuhn as having waffled, or as having "sat on the fence," regarding the question of realism vs. antirealism in science. Reading *SSR*, one gets the impression that Kuhn is irritated—certainly uncomfortable, at least—by this entire clutch of metaphysical issues, and that he prefers to change the subject whenever they come up for discussion. At one point he makes the puzzling remark that traditional scientific realism—the view that different paradigms and theories are about a common, relatively fixed subject matter—"can be neither all wrong nor a mere mistake." He is a torn man. A streak of commonsense in him pulls him toward realism, but an appreciation for the murky details of the history of science pulls him toward antirealism. In later writings Kuhn seems to be fond of the view that the "raw nerve stimuli" impinging on the sensory organs of scientists is basically the same from paradigm to paradigm, from generation to generation, but that such commonly shared sensory experiences encode information so crude and unrefined as to be uninteresting and theoretically useless until filtered through paradigms; but, filtration through a paradigm changes the nature of the original information in a way that creates problems for someone who wants to maintain a form of simple, traditional scientific realism. Kuhn wants to relativize ontology to a paradigm. Questions about what there really is in the universe only make sense when asked "internally," that is, when asked with respect to a paradigm which could provide answers to the questions. Asked as a transparadigmatic question—as a question about which paradigm is the most accurate—no clear answer is forthcoming from Kuhn. On the one hand, he claims that he believes that, in general, later paradigms are better than earlier ones at solving the puzzles that constitute normal science. Yet, on the other hand, he claims that the puzzles can change with change of paradigms, so the first claim is rendered hollow: In what sense is later puzzle solving "better" if we cannot be sure they are the same (or roughly the same) puzzles being solved under both the earlier and later paradigms? In a Postscript to *SSR*, appended to the second edition of the book in 1970, Kuhn wrote perhaps the clearest indication of his view of these matters:

> There is, I think, no theory-independent way to reconstruct phrases like "really there"; the notion of a match between the ontology of a theory and its "real" counterpart in nature now seems to me illusive in principle. Besides, as a historian, I am impressed with the implausibility of the view. (*SSR*, p. 206)

Notice that Kuhn's reported remarks are couched in terms of theories, not in terms of paradigms; again, this illustrates his view that ontological issues only have a well-defined sense *within* a paradigm-based theoretical context—not transtheoretically. Perhaps Kuhn's position can be put this way: The existence or nonexistence of theoretically characterized entities, the truth or falsity of theoretical claims, and the accuracy or inaccuracy of causal laws of nature are all well-defined issues when considered *intra*theoretically, but they are not well-defined issues when considered *inter*theoretically. And, because theories are themselves typically constituent components of a practice-directing paradigm, the same questions are doubly ill-defined when asked in an interparadigmatic sense. If this is a fair assessment of Kuhn's position, then we meet with an interesting result. The position I have described is a form of what Hilary Putnam called **internal realism**, as opposed to the more traditional kind of realism, which Putnam dubbed **metaphysical realism**. Putnam's distinction between these allegedly different kinds of realism is itself highly controversial (as the reader will have noted from my use of the term 'allegedly'), and discussing the complicated differences between them would take us too far offtrack. The glossary of this text contains an entry on each kind of realism, and the reader is referred to the glossary for further information. The important point here is that, if we grant the distinction between internal realism and metaphysical realism for the sake of argument, then Kuhn's considered view does not amount to the sort of across-the-board, anything-goes relativism and antirealism that his more misguided disciples have spread across the intellectual and academic landscape in his name. His considered view does not, for example, imply the result that "truth and falsity don't exist," which Kuhn himself painfully reports was how one group of social scientists misinterpreted him (over his own protests, even—they told him that he didn't understand the full implications of his own masterpiece, while they supposedly did) in the heady days immediately following the publication of *SSR*.

On the other hand, if the most robust kind of realism a Kuhnian can consistently advocate is internal realism, then certainly the thorny problem of paradigm noncorrespondence remains. Incrementalism is a dead horse also for Kuhn; and despite his own words, I do not see how Kuhn can justify any claim that later paradigms are superior at puzzle solving over earlier paradigms; for he needs a notion of common puzzles to solve, and that would require that different paradigms be at least partially commensurable. In summary, there would seem to be no principled, nonbiased way of ranking distinct paradigms, and therefore there can be only subjectively based estimates of progress in science. Accordingly, issues of metaphysical realism—the old-fashioned, traditional kind of realism—go by the board in the Kuhnian model of science.

Perhaps the reader is by now a bit exhausted from the abstract level at which the discussion of Kuhn's model of science has taken place. To lessen that exhaustion and breathe some real-life color into the model, in the next section we will look at the history of modern immunology as a case illustration of the Kuhnian model of science.

7.7 The Case of Modern Immunology

Like any field of science, modern immunology arose out of a misty preparadigmatic period of disorganization and confusion. For most of its prehistory the very notion of an immune system did not exist among practicing physicians. To the ancients and medievals in the Western World infectious illnesses were presumed to be due to some imbalance of the four "humors" (blood, phlegm, yellow bile, and black bile). This was medicine at its most primitive, reduced to the administration of folk remedies for which no controlled experimental studies had been done to establish causal effectiveness, and for which a vast parade of speculative explanatory theories were proposed but never tested. Occasionally, through the dim haze of ignorance and guesswork, a genuine achievement appeared, such as the success at producing immunity to smallpox in humans through intranasal inoculation with the pus of smallpox patients in the first half of the eighteenth century. A few years later, Edward Jenner successfully produced immunity to smallpox through inoculation with pus from the related cowpox infection. These achievements, however, remained for the most part theoretically mysterious. They were "empiric" only; they worked but nobody really had a solid idea why they worked. Their working was a fact among many others that a hobbyist in the field could collect if so desired. Wild speculation blossomed in the attempt to systematize the growing body of unstructured facts about the responses to infection by the human body. In the late nineteenth century Louis Pasteur successfully vaccinated chickens against chicken cholera by inoculation with attenuated chicken cholera organisms; for, by that time, the new field of bacteriology had developed and practitioners were aware of the microscopic world of bacterial organisms and the role such organisms played in certain infectious diseases. Thus, at first immunology was not distinct from bacteriology, the science that studies microscopic bacteria, but a mere branch of it. This should not be surprising, given that the causal mechanisms whereby the immune system performs its functions were utterly unknown before the twentieth century. Instead, the field at first concentrated on the presumed causes of infectious diseases, on the newly discovered world of living microorganisms. Just as we would expect, given Kuhn's model of the evolution of science, competing schools of fact-gathering natural history arose, each trying somehow to patch together a theoretical underpinning that would organize all the facts gathered: the Cellular School and the Humoral School were the main contenders. The Cellular School concentrated its attention on discoveries showing the existence of different natural kinds of cells in the bodies of mammals whose functions appeared to be related to immune protection of the organism, cells such as phagocytes that physically engulf foreign bacteria and digest them. The cellularists argued that all immune phenomena were cellular in nature. The Humoral School, in contrast, concentrated on growing evidence that the blood serum of mammals contained various soluble (noncellular) biochemical substances (what we now call antibodies) whose functions appeared to be related to immune protection of the organism. The humoralists argued that all immune phenomena were ultimately humoral in nature. The leading speculative theorists split into opposing camps. Elie Metchnikov, the future Nobel laureate in medicine (1908), along with most of the researchers based at Pasteur's Institute in Paris, was a devoted advocate of the Cellular School, while researchers based at the research institute run by Robert Koch, another future Nobel laureate in medicine (1905), in Germany, belonged to the Humoral School. Each school fought for dominance over the other. Research tended to focus on establishing the fundamental principles governing immune phenomena rather than on ferreting out minute microstructural facets of immune phenomena. Each school took pains to produce and then point out experimental results that were seemingly incompatible with the opposing school's methodological orientation, speculative doctrine, or explanatory theories. For example, the humoralists produced experimental evidence that immunity could be induced in an organism in the absence of any cells at all, the so-called Pfeiffer Phenomenon. In their defense the cellularists pointed out experimental support for their model of inflammation (they claimed it was due to the biochemical aftermath of phagocytic engulfment of foreign organisms), a subject matter on which the humoral theory at that point had little of interest to say.

Into this war of schools stepped Paul Ehrlich, the Isaac Newton of modern immunology, during the closing decade of the nineteenth century. Ehrlich was based at Koch's institute in Berlin. At the time Ehrlich first began to work on the nature of immune phenomena it is important to remind ourselves that not even the basic vocabulary of immunology existed. The terms 'antibody', antigen', 'complement', 'receptor', 'T cell', and so on, either had not been coined and were not available to practitioners to help them organize the rapidly coalescing mass of data on immune functioning, or else they had been coined but their usages were neither widespread nor standardized. These specialized terms awaited a paradigm in which they could be given a home, in which their coinage as a mutually interdependent network of terms made sense, and in which their uses could be standardized in a stable way so as to render them useful for explanatory and predictive purposes. Ehrlich was the practitioner who would be responsible for the introduction of the first paradigm in immunology: the Ehrlich **side-chain model** of immune functioning. What Ehrlich did is in retrospect all too obvious: He combined what was best in both the preceding two schools of speculation about immune functioning. The result was the publication in 1897 of his famous side-chain model of the generation of what we would now call antibodies, a model later clarified, put into its classic form, and delivered as his Croonian Lectures of 1900. The essential basics of the side-chain theory are rather easy to understand. First, antibodies (although that term was not yet in widespread use, Ehrlich called some side-chain molecules *Zwischenkörper*, 'between bodies') are natural products of cells; cells churn them out as part of their normal physiological behavior. These antibodies are transported somehow out to the outer surfaces of the cells that produce them, where they lock in place as "side-chains" sticking out off the surface. Each side-chain is presumably structurally different. Second, the antigen, the substance that elicits the immune response, comes into contact with the cell, and, if the structure of the antigen "fits" the structure of the side-chain antibody, the former locks onto the latter, in something like the way a key fits into a lock. Third, through some unknown process, this triggers the cell into producing many more molecules of the side-chain antibody and subsequently "shedding" the excess molecules of the antibody into the general blood circulation where they are available to destroy the antigen in question should they meet with it.

Here was an elegant and fruitful model, an achievement of profoundly wide-ranging impact, and which, with certain notable emendations has been largely verified by subsequent generations of research. It constituted a Kuhnian paradigm, an achievement sufficiently rich to settle fundamentals and open up a period of normal science research into structural details. After the side-chain model swept through the community of practitioners, immunology entered its first post-paradigmatic period of normal science, a period during which a tremendous amount of detailed research made possible an investigation of immune phenomena at a level of detail unimaginable during the time when basic fundamentals were still up for grabs.

For example, after the introduction of the side-chain model practitioners were able to focus on figuring out the minute structural details of antibody specificity, that feature of each antibody that allows it to have a remarkably precise specificity for what it is capable of attacking. Your typical antibody attacks almost nothing else except its instigating antigen, and cross-reactions with structurally similar antigens are quite narrow rather than wide. Here was just the sort of puzzle that Kuhn claimed it is the business of normal science to solve. Here was just the kind of secure research—research which does not challenge the reigning paradigm but rather takes it for granted—in which a definite solution was virtually guaranteed, if only the practitioners were inventive and patient enough to ferret it out from nature's hidden recesses.

To be sure, the side-chain theory had its enemies; yet, overall, immunology came of age as a genuine experimental research science during the first three decades of the twentieth century primarily as a result of Ehrlich's paradigm serving as the directing force in the field. If you could do it the way Ehrlich had done it in his side-chain model, why that was successful science in immunology. In the decade of 1900 to 1910, most of the basic terminology of modern immunological science was standardized, although some variance due to local doctrinaire disputes continued until World War II. In particular, much has been made by historians of medicine about the dispute between the Ehrlichians and the anti-Ehrlichians, especially over the issue of autoimmunity.

Recall that Kuhn insisted upon the essential incompleteness of paradigms. No paradigm is perfect. No paradigm can handle every research result that comes to light under its direction. Every paradigm must confront anomalies, research results that do not seem compatible with it. Ehrlich's side-chain paradigm was no exception to this principle. The exact details of just how the antigen fit onto the side-chains, just how the side-chains were formed inside cells, which cells so formed them, how the side-chains were transported out to the exterior surfaces of the cells that made them, whether there was only one or more than one binding site per side-chain molecule, how the shedding of excess side-chains took place, how the cell was signaled to produce excess molecules of the side-chain it made—all these and many more issues remained to be puzzled over by practitioners. It was to be expected that as more and more detailed observational data were slowly collected, anomalies would begin to appear. One claim on which the side-chain model seemed to be incorrect was its insistence (actually, Ehrlich's insistence) that the binding of antigen to antibody (side-chain) was biochemically irreversible. Critics hostile to the side-chain theory produced experimental evidence early on that the binding of antigen to antibody was reversible. On the other hand, the side-chain model was correct in insisting that the binding of antibody to antigen was chemical in nature rather than absorptive or colloidal, as some of the critics surmised. Notice how minutely detailed these disputes are. Probably, the reader is not familiar enough with the details of microchemistry even to be sure what 'colloidal' means. This is one result of paradigm-based normal science research: It scales down disputes from being about fundamentals to being about detailed minutiae.

One anomaly that arose quite early was the appearance of evidence that the immune system could mount an antibody attack on certain parts of the organism's own tissues and fluids. The breakthrough research on this question was done by the future Nobel laureate in medicine (1930) Karl Landsteiner, who along with Julius Donath discovered in 1904 the first human disease definitely based on the production of so-called *autoantibodies*, antibodies to the patient's own tissues or fluids: paroxysmal cold hemoglobinuria, a disease in which the patient's immune system produces antibodies to certain antigens on the surfaces of the patient's own red blood cells. The antibodies attack the red blood cells only within a narrow temperature range, only in the "cold." This sort of situation was anomalous because one apparent implication of the side-chain paradigm was that the organism's cells would not produce side-chains capable of attacking its own flesh—Ehrlich declared that such a phenomenon would be "exceedingly dysteleologic," that is, very much nonpurposeful. The phrase used to denote this alleged natural inability of organisms to launch immune self-attacks was horror autotoxicus: Nature abhors creatures capable of sustaining self-toxic physiological processes, and such processes do not evolve as natural components in the physiology of organisms.

Another anomalous phenomenon was the growing consensus in other research quarters that the immune system was capable of launching an attack on seemingly harmless targets. In chapter 1 we reviewed the history of the discovery of allergic phenomena in some detail, and I shall not repeat it here. Again, nothing in the side-chain paradigm would have suggested such a phenomenon as allergy. From an evolutionary perspective, wasting precious physiological resources to attack innocuous targets such as pollen grains and mold spores would seem to be just as "exceedingly dysteleologic" as outright autoimmune attack. Autoimmunity and allergy were anomalous with respect to the side-chain paradigm; yet the reader should be advised that the paradigm hardly fell just because it was confronted with those two odd phenomena. On the contrary, normal research under the side-chain paradigm rolled on unchecked in the face of autoimmunity and allergy. For, as Kuhn indicated, the assumption of the normal science practitioner is that, if only research is ingenious enough, if only it is skillful enough, the reigning paradigm can be made to accommodate the seemingly anomalous data. The hope was that somehow, someway, eventually autoimmunity and allergy could be accomodated by the side-chain model of immunological phenomena. Historians of medicine have noted that between 1900 and 1950 immunology was dominated by physicochemical research, not biological/medical research. The general orientation was chemical not biological, and so biological anomalies got put on hold while the research "stars" concentrated on working out the minute details of the antigen-antibody chemical bonding and specificity. Only after about 1950 did a significant change take place, a change that saw biological issues once again come to the front of the stage. A generation arose that was bothered by the accumulated mass of anomalous data the old side-chain model could not accommodate (except perhaps by *ad hoc* adjustments that left the model unsystematic and predictively imprecise). The result was a revolution of sorts within immunology, a revolution that culminated in the clonal selection theory of the late 1950s, whose history we have already reviewed in the previous chapter. Here was a new paradigm, more powerful than the old side-chain model in its ability to accommodate such phenomena as autoimmunity and allergy. A new period of normal science ensued in the 1960s, a period of normal research directed by the clonal selection paradigm in which different puzzles now had to be solved. Indeed, one notable difference between the new paradigm and the preceding one was that autoimmunity was almost to be expected under the clonal selection theory, for that theory suggested how it could likely happen that the fragile control mechanisms in the immune system could be easily put into imbalance and disarray. What the clonal selection theory singled out for puzzle-solving research that the Ehrlichian side-chain theory had not singled out for such research was the whole business of immune system *control*: how the different branches of the immune system (for example, cellular vs. humoral, B cell vs. T cell, phagocytic vs. cytotoxic) interact in concert and against each other, how autoimmune clones are suppressed or unsuppressed to produce autoimmune disease, how the organism "identifies" what to attack and what to leave alone (the problem of "self vs. nonself"). The last issue was hardly noticed under the old side-chain paradigm, for that paradigm had not suggested that one of the immune system's jobs was to differentiate self tissues and fluids from everything else—such a research puzzle simply would not have occurred, and did not occur, during the reign of the side-chain paradigm in immunology. Illustrated here is the Kuhnian point that a change of paradigm in a field can result in a reorientation of what the field is about, what its chief focus is, and what puzzles its practitioners find of intrinsic interest to work on.

But consider the more metaphysical aspects of the Kuhnian model of science. Are the clonal selection and side-chain paradigms really incommensurable—not comparable against each other as to their respective and relative merits? Was the advance of knowledge in immunology discontinuous, not cumulative, across the change of paradigms? Does the history of modern immunology support the Kuhnian contention that the "facts" in a field are, at least partially, constructed by the methods of practice used in that field, rather than discovered as they are "in themselves"? Are an Ehrlichian and a clonal selectionist talking about different entities when they each use a term common to both paradigms such as 'antibody'?

My view is that the answer to all four of the above questions is no. But these days such a realist position in the philosophy of science is growing increasingly unfashionable among many thinkers. On the contrary, Kuhn's more radical metaphysical claims are often taken as the new Holy Writ in fields of scholarship that see themselves as qualified to give a general critique of science. In the next chapter we take a closer look at this reemergence of the realism/antirealism dispute in post-Kuhnian philosophy of science.

Further Readings

The watershed work by Thomas Kuhn is, of course, *The Structure of Scientific Revolutions*, 2nd. ed., enlarged, Chicago, University of Chicago Press, 1970. For further development by Kuhn of his original views see also, "Second Thoughts on Paradigms," in *The Structure of Scientific Theories*, 2nd. ed., edited by F. Suppe, Urbana, Ill., University of Illinois Press, 1977, pp. 459–82, and "Commensurability, Comparability, Communicability," in *PSA 1982 volume 2*, edited by P. Asquith and T. Nickles, East Lansing, Mich., Philosophy of Science Association, 1982, pp. 669–88. A detailed history of the dispute between Cellularists and Humoralists during the early years of modern immunology can be found in Arthur Silverstein's *A History of Immunology*, San Diego, Academic Press, 1989, pp. 38–58.

8

The Social Constructivist Challenge

The demise of the positivist model of science and the rise of Kuhnian historicism ushered in a momentous change in philosophy of science. The previous consensus view that science, at least on occasion, discovered a preexisting objective reality gave way to alternative critical accounts of science that stripped it of any such achievement. These new critical accounts were invariably relativist in epistemology, and, for the most part, antirealist in ontology—but the relativism and antirealism in question were not the traditional philosophical brands of relativism and antirealism. The new forms of relativism and antirealism were born within the nest of the social sciences. Many Kuhn-inspired critics of science took Kuhn's ultimate point to be that the philosophy of science, as traditionally conceived since the time of Immanuel Kant, was dead. Its old job would be taken over by a successor discipline called variously the sociology of science, the sociology of knowledge, or science and technology studies. And this new discipline would not go about the old job in the old way. One leading spokesperson of this new discipline is the sociologist David Bloor, who published a treatise outlining the details of the new successor discipline to philosophy of science. Bloor called this new successor discipline the sociology of knowledge, and the particular type of it that he advocated he called the **Strong Programme**.

8.1 The Strong Programme

The Strong Programme was born partly out of a reaction to traditional philosophers of science who sought a compromise between Kuhn and traditional neopositivist philosophies of science. With Kuhn, these philosophers agreed (rather grudgingly) that different paradigms are incommensurable. But, unlike Kuhn, they argued that such incommensurability did not cut us off from giving a *rational* assessment of competing paradigms: for it is possible to judge paradigms by how well or poorly they solve the research puzzles they set for themselves, in their own terms, you might say. The upshot of this compromise model of science was that *non*rational "social" factors in the practice of science were to be called on *only* to explain failure and error within a paradigm-based research tradition. Wherever a paradigm-based form of inquiry met with success, however, the explanation for that success was that the paradigm was accurate, that the model of reality it contained was true. Thus, the compromise position amounted to the claim that nonrational social causes are to be appealed to only when explaining scientific error, never in explaining scientific success, thereby securing the essential rationality of science even in the face of major error.

This asymmetry is unacceptable to Bloor and his fellow sociologists of knowledge. Against it, the Strong Programme holds that false beliefs and true beliefs in science are to be explained by the same kinds of social, nonrational causes. That is, in the Strong Programme one is never allowed to explain why practitioners of a science hold a particular belief by saying "because that belief is true." This is the source of the name "Strong Programme"; for the "weak" programme in sociology of science is the one advocated in the compromise view, in which only erroneous beliefs receive explanations in terms of nonrational "interference" factors. On the contrary, Bloor insists, being true of reality does not explain why a true belief comes to be held by scientists. The usual reply by philosophers of science is that, if truth does not serve, in any capacity whatsoever, as a constraint on theoretical belief in science, then the content of science would surely float free of any theory-independent reality "out there"—so free of it that the content of science in the end wouldn't be *about* that theory-independent reality at all, but about something else. Hence, Bloor's Strong Programme cannot avoid either antirealism or an "anything goes" form of relativism.

Bloor is aware of the above charge and he replies to it that the Strong Programme is not necessarily committed to antirealism; nor, for that matter, is it necessarily committed to an "anything goes" form of relativism. He does admit that the Strong Programme is a relativist model of science, but it's a brand of relativism of a rather modest sort (that is, less than an "anything goes" relativism). These are claims that a philosopher of science ought to be duly suspicious of, and we shall be very insistent ourselves that Bloor make good on them. The reason for our suspicion is that the Strong Programme, as set out by Bloor, aims for the outcome of showing that *scientific knowledge is socially constructed rather than discovered*; and, any time somebody aims for that outcome, the worry arises that antirealism and an "anything goes" version of relativism are going to crash the party uninvited despite the host's best efforts to bar the door against them.

Bloor claims that the Strong Programme is not committed to antirealism because a sociologist of knowledge working in accord with the Strong Programme certainly admits the existence of a practice-independent and theory-independent material world. No advocate of the Strong Programme, says Bloor, is an idealist. What the Strong Programme asserts is that the nature and structure of the theory-independent material world is so filtered through "socially sanctioned metaphors" that its nature and structure do not significantly effect the content of scientific theories. The idea seems to be that the practice-independent world is a kind of unknowable something-or-other we are cut off from any direct contact with; hence, science is not about this world, it is about this world under descriptions and explanations licensed by contingent and conventional social metaphors and relationships. It follows that the reality that science reveals is not the world's reality, it is the reality of our social forms of life. In the years since Bloor's manifesto appeared, sociologists of knowledge have done a great deal of work producing case studies designed to show how the content of even true beliefs in science is determined, not by evidence regarding the structure of the independent reality under investigation, but by nonrational social factors such as the political commitments of the practitioners in question, the unconscious projection of social relationships onto nature, desire for fame, etc. This sets up an argument that Bloor does not actually give, but which I claim is a fair representation of the Strong Programme:

1. True beliefs in science are just as caused by the interpretation of data via projected social metaphors as are false beliefs

therefore

2. The content of scientific beliefs is as socially determined as anything else, which is to say it is more socially determined than it is nonsocially determined (the contrast here is with determination by what is the case about a theory-independent reality)

therefore

3. Scientific knowledge is socially constructed rather than discovered.

The central problem with the Strong Programme is that Bloor, despite his best efforts to the contrary, cannot prevent either antirealism or an "anything goes" form of relativism from crashing the party. To see why, let us take realism to be indicated in the last remark of the following conversation:

4. *Q:* Why does scientist S hold belief B?
5. *A:* Because holding B meets with success in practice.
6. *Q:* Why does holding B meet with success in practice?
7. *A:* Because B is true (of a theory-independent reality).

Answer 7 is the realist answer to question 6. The Bloorian social constructivist, however, is prevented from giving answer 7 by the rule in the Strong Programme that truth is never an explanation for why a belief is held. On the contrary, the Strong Programme must substitute a different answer to question 6:

7′. Because current social metaphors license a verdict of success with respect to using belief B in practice.

But 7′ renders the Strong Programme vulnerable, I claim, to an "anything goes" form of relativism by the following argument:

8. Success verdicts are licensed by particular social metaphors and relationships that are projected onto nature in scientific method and theory.
9. There are no objectively based limits on which social metaphors and relationships could come to dominate a culture.

therefore

10. In principle, anything could come to go in science.

Bloor would surely object to the conclusion directly on the grounds that science must meet with some minimum level of success at manipulating the material world per se; hence, it is just plain false that anything could come to go in science. But the previous argument (4–7), I claim, cuts off this response, for that response helps itself to something Bloor is not entitled to given his other claims: a social-metaphor-independent notion of success at manipulating the material, theory-independent world.

I am arguing that the Strong Programme is not entitled to use the idea of success at interacting with a practice-independent reality to escape "anything goes" relativism because such success will be just one more socially determined and heavily filtered judgment, not about reality, but about reality under socially sanctioned description. The problem, as I see it, is the same one we've had now for almost 200 years since Kant: If one is serious about the Kantian distinction between **phenomena** and **noumena**, between things-as-interacted-with-via-social-metaphors and things-in-themselves, then, in the end, one cannot escape a completely skeptical antirealism about things-in-themselves, or complete relativism concerning our theories of things-in-themselves. If you conceive of the theory-independent world as a remote and unknowable thing-in-itself, then of course you will be led to the conclusion that we can know nothing of it, and its nature and structure therefore cannot significantly constrain the content of our knowledge, scientific or otherwise. Bloor defines true beliefs to be those that work without realizing that to escape antirealist relativism he shall need a social-metaphor-independent notion of "working." But such a notion is not available to him, for verdicts about which beliefs work and which ones don't work are as much under the sway of social metaphors that float free of the theory-independent reality as anything else.

We should note the likely fact that Bloor would object strongly to premise 9. Bloor's view is that the Strong Programme is the one model of science that can honor science's objectivity, because it locates that objectivity in the proper place: the social. Bloor holds that *the objective is the social*, and it couldn't be anything else, because the theory-independent world is an unknowable thing-in-itself; hence, the only source of intersubjective agreement and theoretical stability is society itself. This is a fuzzy argument, and one senses that the term 'social' is being so stretched out to cover so many different kinds of phenomena as to bleed it of all significance. Bloor's strategy is to claim that only the social world could provide the kind of stability over time that underwrites intersubjective agreement, and only intersubjective agreement could underwrite objectivity. Premise 9 is false, Bloor would claim, because no one person or group of persons can alter at will the social metaphors and relationships that form the structural components of society. Not just any social metaphor could be set up to license success verdicts, seems to be Bloor's suggestion. Society is bigger than any elements within it, and it has a dynamic of its own independent of the personal psychological desires of its members. Will this response by Bloor work? It doesn't seem very likely that it will. Nothing in the case studies he presents suggests that there are fixed limits on the different forms social metaphors and relationships can take in human life, on which ones can come to dominate a culture. For example, one might have thought that reason itself would supply fixed limits, that the fundamental principles of logic and mathematics would simply rule out certain forms of social metaphor and practice on the grounds that those social metaphors and practices are incoherent, inconsistent, or absurd. Unfortunately for Bloor, he argues the very opposite. He insists that not even logic and mathematics contain fixed truths that could supply fixed limits to the forms of social evolution among human beings. A closer look at what he says about this particular issue is instructive.

The central claim in the Strong Programme is that the content of scientific theories is not significantly affected by the nature and structure of the theory-independent reality that is nevertheless, in some Kantian fashion, "out there." This claim must be filled in for specific cases of successful science. In other words, even in positive cases of success (rather than negative cases of error), the content of the theories in question is determined by nonrational social factors—not by "the reality" the theories are incorrectly thought to be theories about. Bloor attempts to illustrate this central claim with a number of case studies from, surprisingly, the science of mathematics. Unfortunately, there is a tendency for his case studies not to show what he thinks they show. *Instead of illustrating that be-*

lief content is a function of projected social metaphors, what Bloor's case illustrations show is that the Quine-Duhem Thesis is alive and working well in science. Bloor takes cases in which practitioners respond to an anomaly by adjusting other background beliefs—rather than by adjusting the foreground test belief—to show that it is not the alleged theory-independent reality under study that is constraining the adjustment, but rather social factors are the source of the adjustment. But these cases of adjustments made to background beliefs in order to deal with anomalies need not be taken to show any such thing; for they can be seen more accurately as cases in which the adjustments in question were *in part* driven by what is the case about a practice-independent reality under investigation. It is an open question what drives such adjustments. To take an example Bloor investigates extensively, the ancient Greeks took the "proof" that the square root of 2 cannot be represented as a fraction composed of integers to be a proof that the square root of 2 is not a number at all, whereas moderns take it as a proof that there are certain numbers that are numbers, but "irrational" numbers. The ancients held that there is only one kind of number and the square root of 2 isn't among that one kind. The moderns believe that there are multiple kinds of numbers and the square root of 2 is a member of one of the more exotic kinds. Rather than treating this case as an example of an adjustment to background beliefs that counts as progress in ferreting out the structure of a reality which is independent of mathematical inquiry—that counts as a case in which moderns discovered something the ancients overlooked—Bloor interprets it as showing that there is no social-metaphor-independent fact of the matter whether the square root of 2 is a number or is not a number. I put the word 'proof' in quotes a few sentences back because one lesson Bloor draws from this case is that the notion of proof itself is a social creation. Proofs, he says, have no intrinsic meanings. To call something a proof is a "moral" act, it is to place a social approval upon it and take it to establish a piece of knowledge. The intrinsic ambiguity of formal demonstrations renders the force of logic and reason entirely reactive to cultural and social interpretations. One problem here is that Bloor fails to appreciate the fact that, if he is right, it strengthens even more premise 9 (see page 160). If even formal proofs in logic and mathematics can be reinterpreted to mean something other than what they are thought to mean, then "anything really could come to go," at least in principle, for not even math and logic can serve as universal and absolute constraints on our belief formation or behavior. Another problem with Bloor's argument has a more systematic character. The case of mathematical proof might be rather atypical. Traditionally, philosophers have given mathematics a special epistemological status with respect to the rest of our empirical inquiry. A case could be made independently of specifically social constructivist assumptions that mathematics involves a large dose of conventionality—that mathematical concepts are especially conventional to begin with—a matter of how we agree to agree on the meanings of special symbols and procedures. If that is so, then Bloor has no business generalizing from the case of mathematics to other domains of inquiry where things may not be as convention based. It is difficult to imagine Bloor being able to make the same sort of argument about proof in, say, immunology, where there is presumably (or so says the realist) a wealth of nonconvention-based physical structure that impacts on what counts as conclusive evidence for what else.

8.2 Roll Over Socrates: Philosophy as the Bad Guy

The Strong Programme is beginning to look like a Trojan horse inside the walled City of Science. Bloor promises in his argument for the programme that it will neither damage, cheapen, nor insult science, but instead it will place science on the only proper footing on which its objectivity could be legitimately respected. What we have seen of the Strong Programme so far, however, suggests that it is basically Kantian idealism dressed up in the latest social science clothing, Kantian idealism removed from the individual mind (which is what Kant cared about) and placed in some sort of collective social mind.

The writings of the Strong Programme advocates are unfortunately filled with mishandled case studies such as the one above about the proof that the square root of 2 is a nonrepeating decimal number. It is perhaps not surprising that such cases are mishandled by proponents of the Strong Programme, for one notable feature of the sociology of science is the extent to which it depends on either redefined terms or vaguely defined terms that otherwise have a relatively clear sense within philosophical discourse. Early in his treatise, for example, Bloor attacks traditional epistemology as it is practiced by philosophers. He announces that the sociology of knowledge takes the term 'knowledge' to refer not to beliefs that are justified and true, or reliably produced and true, or any of the other characterizations currently offered by philosophers, but rather the term refers to whatever beliefs are "socially endorsed" by a given community. Truth is not a necessary condition of knowledge, Bloor boldly asserts. So, if the belief that the square root of 2 is not a number at all is a socially endorsed belief in a given culture, then the members of that culture can be said to *know* that the square root of 2 is not a number. This virtually assures relativism by definition, and Bloor defends this move by suggesting that the traditional definitions of knowledge in philosophical epistemology beg the question against relativism by assuring an absolutist form of knowledge. In fact, Bloor is refreshingly honest about this sort of polemical aspect to the sociology of knowledge. He admits that the sociology of knowledge, certainly as it is embodied in the Strong Programme, is a competitor discipline to philosophy of science and to philosophical epistemology. They compete on the same intellectual turf, and the sociology of science wins out over philosophy of science and philosophical epistemology, in Bloor's view. A philosopher is pretty much at a loss what to say about Bloor's redefinition of knowledge to mean something like popular belief. To amend a phrase from Chuck Berry's "Roll Over Beethoven," Bloor's message to philosophy might be said to be, "Roll over Socrates!" Any competent student of philosophy is well aware of how intensely and successfully that mighty stonecutter of ancient Athens argued for the distinction between knowledge and popular opinion.

It is clear from his writings that Bloor understands the Strong Programme in the sociology of knowledge to be a form of **naturalism**. As such, the programme is opposed to what Bloor considers the nonnaturalism of traditional philosophical epistemology. What Bloor wants to show is that knowledge, truth, and other epistemological concepts can be given perfectly natural, causal characterizations in terms of contingent, socially determined explanatory metaphors, and that therefore there is no need to posit an abstract "world" where eternally abiding logical entailments, relations, and conceptual truths "reside" and thereby fix in place an objective epistemological reality. Certainly he is correct that much of traditional philosophy has opted for such an otherworldly treatment of abstract concepts like truth, knowledge, and analytic entailment. But not all philosophers stick with the traditional party line on these matters. Quine himself railed against thinking of meanings and concepts as quasi-objects in some heaven of abstract entities where they hover about with timelessly fixed attributes that our linguistic practices here on earth must respect by matching correctly. Bloor is a devout antiessentialist, in the jargon of contemporary philosophy, and therefore he seeks a model of knowledge and science in which neither involves the uncovering of the supposedly fixed essences of things. He believes that specifically philosophical accounts of these matters are hopelessly essentialist, therefore they are all bogus accounts. Philosophical accounts of concepts seek to mystify them, in Bloor's view. He means by this that philosophical accounts seek to portray the concept under analysis to be one so mysteriously abstract and remote from ordinary causal and material life as to be inaccessible to anyone but the specially initiated professional philosopher. This will ensure, it seems, that the rest of the world will always need philosophers around to explain these mystifying notions only they understand. Professional philosophy is thus mostly stripped of its cognitive merit and debunked as being a form of self-serving intellectualism.

Against the mystification of knowledge by philosophy the sociology of knowledge supposedly stands as a bastion of intellectual egalitarianism, as the champion of the common person who lives in the everyday social world and fights the good fight against the goofy philosophers addicted to their otherworldly Realm of the Conceptual. If a social constructivist movie was made of human intellectual inquiry, philosophy would be like the proverbial "bad guy" of an American Western, the one with the skinny mustache, the squinty eyes, the bad skin, and the knurled hands. But ordinary folk terrified by the sour-souled philosophers should take heart. The sociology of knowledge has arrived in town for the showdown, and you can bet the house that it will be the evil mystifiers of human knowledge who shall eat the dust after the guns blaze at high noon. For example, according to Bloor, philosophers invented the notion of *necessity* so that they could succeed in mystifying concepts like causation and logical truth. In fact, argues Bloor, necessity is actually nothing more than a projection onto nature of a central social metaphor in all human societies, *moral authority*. To claim that such-and-such a statement is a logically necessary truth, or that such-and-such a sequence of events was causally necessary, is merely to invest that sentence and that sequence of events with the sort of authority over human affairs that moral taboos and commands have. The laws of a science, for example, do not represent essential connections between structural aspects of some inquiry-independent reality, rather they separate out certain patterns within our practices in the area of inquiry in question and invest them with the power of moral authority. These are wonderfully clear examples of how the Strong Programme supposedly debunks and demystifies scientific concepts that philosophers of science allegedly insist on treating in some more mystifying and essentialist way.

One wants to ask the obvious question: If necessity is a projection onto nature of the social metaphor of moral authority, then because it is a mere *projection* by us, does this mean that there are no morally authoritative necessities "out there" in the inanimate material world? Bloor seems strangely cavalier about this possible implication of his argument. A projection is imposed from outside, and it can be a distortion of what it is imposed on or else it can fit it perfectly. In science, among other things, we want to know what must be the case about things, if anything must be the case about them. Apparently, we cannot know that, according to Bloor, for the most we can know is what the projected moral authority tells us we think must be the case about things, and then only in the sense of what we think *ought* to be the case about them. The reader would do well to file this result away in a safe place, for the feminist social constructivists whom we will investigate in the next chapter will seize on this aspect of social constructivism and make it one of the centerpieces of their critiques of science.

8.3 The Case Method with a Vengeance

The sociology of knowledge is an empirical discipline. It will stand or fall on whether its pronouncements are consistent with the relevant data. The relevant data are case studies in the history and/or practice of science. It is no wonder then that social constructivist models of science are heavily dependent—perhaps even more so than Kuhn himself was—on case studies of scientific practice and method. Bruno Latour and Steve Woolgar are two prominent social constructivists who are specialists in the right sort of case studies. The idea is to do the case studies, not the way a historian would, but the way an anthropologist doing field work would. This is what Latour did when he showed up at a research laboratory housed in the Salk Institute in California, a laboratory specializing in the investigation of hormones that originate in the nervous system. He wanted to observe practitioners working in the lab in the hope that they might follow the social construction of a scientific "fact" from beginning to end. Woolgar later joined Latour to coauthor a study based on Latour's observations. They wrote with a heavy dose of the self-referential plural despite Woolgar's absence from the lab. My own remarks below will respect their preference in this matter by discussing their study as if it was based on a jointly experienced sociological field exercise.

Latour and Woolgar admit to approaching science with an attitude not unlike that with which field anthropologists approach the study of a primitive and alien culture. Science to them is alien in the same way a previously unknown tribal culture is alien. Further, just as no well-trained anthropologist would take the primitive natives' own interpretation of their culture as a correct account of it, so Latour and Woolgar argue that it is not appropriate to take scientists' own interpretations of their work as correct accounts of it. The sociologist of science must avoid "going native," as they put it. On the contrary, science is to be accorded initially no more rationality as a knowledge-gathering method than is, say, primitive witchcraft. Latour and Woolgar come close to confessing outright this sort of debunking view of science as witchcraft in a footnote: "One particularly useful source [for their own study] was Auge's. . . analysis of witchcraft in the Ivory Coast, which provides an intellectual framework for resistance to being impressed by scientific endeavour." (*Laboratory Life*, p. 260). Just as the field anthropologist never really believes that the tribal sorcerer under study is actually breaking the laws of nature with his or her witchcraft, so, according to Latour and Woolgar, the sociologist of science ought never to believe the scientists' own claims that science finds out "the facts," the "actual way things really are." As Latour and Woolgar see it, the task of the sociologist of science is to give a debunking account of scientific practice, one that shows how practitioners of science become deluded into thinking that they are discovering a reality that was already in existence rather than constructing a reality by artificial means (which is actually what they are doing instead). It is no accident therefore when Latour and Woolgar admit unashamedly that they studied the behavior of the scientists in the lab as though those scientists were the postindustrial equivalent of, and here I want to use the very expression Latour and Woolgar use, *primitive sorcerers* (*Laboratory Life*, p. 29) engaged in routinized behavior that was entirely "strange" and virtually incomprehensible to Latour and Woolgar. One bizarre oddity to their study (which the intelligent reader can hardly miss) is how "strange" Latour and Woolgar seem in failing to understand or appreciate the obvious explanation for this strangeness they felt toward what they were observing in the lab: Namely, unfamiliarity and lack of experience with *any* kind of activity will tend to make that activity seem strange when seen from the outside. But instead of condemning their own ignorance as scientifically naive observers, Latour and Woolgar conclude that they are the only people in the lab capable of an objective understanding of what the scientists are doing. Ignorance is a necessary condition of wisdom, apparently, in the ironic world of the sociology of science.

What those scientists are doing is not what those scientists think that they are doing, according to Latour and Woolgar. The field study is rather humorous to read, although some of the humor is surely unintentional on Latour and Woolgar's part. The two field anthropologists are thoroughly perplexed by the daily activities of the lab workers, activities that the sociologists describe in the most observationally literal way they can invent. Machines hum and churn and produce "wavy lines," which are then superimposed; flasks of liquid are shaken, then argued over; workers consult texts in order to produce more "inscriptions." The attempt is to describe moment-to-moment practice in a bioscience laboratory in such a manner as to make it out to be eccentric and arbitrary enough to qualify as irrational: science as a more fancy form of witchcraft. What truly shocks Latour and Woolgar is the claim made by the practitioners they are studying that they, the practitioners, are doing something much more worthwhile than arbitrary witchcraft—that they are discovering objective structures in the domain of nature under their investigation, that they are finding out objective facts about neurohormones. Latour and Woolgar are sure that this is not so, that any claim to have discovered preexisting material realities is a self-congratulatory form of exaggeration on the part of the scientists, basically nothing more than arrogant hubris; for all that is *really* going on in the lab is that, machines hum and churn and produce wavy lines, flasks of colored liquid are shaken, then argued about, workers consult texts in order to produce more inscriptions, and so on.

The problem seems to be that Latour and Woolgar make a basic mistake in the epistemology of science. They take the fact that, often in science, the detection of an objectively existing structural reality requires an elaborate detection procedure (involving elaborate machines and so on) to mean that no such objectively existing structural realities exist; for what the practitioners are "detecting" in such cases is an artificial structural reality, one created by the elaborate detection procedure. In other words, for Latour and Woolgar, The wavy lines produced by the fancy machine that produces wavy lines are not wavy lines *about anything except the very process that produced them*. The content of science is merely the process by which it is artificially constructed as knowledge—it is not the structure of some independent reality. No properties of an independent reality are represented in the wavy lines, for the wavy lines are entirely artificially manufactured by the machine that makes wavy lines. When the scientists themselves object to this that the wavy lines contain peaks and valleys which represent some of the chemical properties of a hormone sample that was put into the machine that makes wavy lines, Latour and Woolgar will believe none of it; for that is too much like listening to the tribal witch doctors insisting that the spell they cast really did cure the sick daughter of the assistant chief by driving away the independently existing evil spirits.

Latour and Woolgar will have none of it in this way because they make a specific philosophical mistake: From the fact that our *being able to say or to claim* that X exists depends on an elaborate and artificial procedure for detecting X, they mistakenly think it follows that X's existence itself depends on that elaborate and artificial detection procedure. But, of course, this is an absurd as well as an invalid inference. To see the absurdity, consider the following implication of the Latour and Woolgar view. We cannot observe viruses without very elaborate and wholly artificial equipment such as electron microscopes, special chemical stains, and so on. If Latour and Woolgar are correct, that would imply that the *existence* of viruses *depends* on such artificial equipment and such elaborate detection procedures. If such procedures and equipment did not exist, *then viruses would not exist.*

Why do Latour and Woolgar conceive the sociology of science to be this kind of debunking project? What motivates their confidence that practitioners of modern science are as serenely mistaken about the real nature of their own business as are the tribal witch doctors about the real nature of their business? Well, Latour and Woolgar claim that the facts—indeed, all the facts—of science are entirely socially constructed. Scientific facts are really about their own social construction, not about some independent reality. But they realize that everybody thinks scientific facts are about some inquiry-independent reality. Now most scientists are fairly smart. What would lead smart people to make this sort of mistake about the real status of scientific facts? One central chore of the sociology of science must be to explain how it is that what is entirely socially constructed and about one sort of subject matter is almost universally mistaken by otherwise bright human beings to be not entirely socially constructed and about a completely different kind of subject matter. This will force the sociology of science to become primarily a project of debunking. Latour and Woolgar attempt to provide the missing explanation of the mistake: The methods of science are designed to "cover up" the social construction of the products of science. Once a purported fact has been established by scientific means, the elaborate and artificial procedures used to establish it conveniently disappear from notice and are forgotten. The fact now seems to "stand on its own"; hence, the delusion (Latour and Woolgar's word for it is 'illusion', but they are being sloppy here—they mean to speak of what is more properly called a delusion) naturally arises that the fact is about an independently existing objective reality and not about its own process of construction through organized social behavior. That, according to Latour and Woolgar, is why the wavy lines produced by the machine that makes wavy lines are universally held to have been made by the interaction of an independent reality with the machine, when actually they were complete creations of the machine and how it was manipulated by its users.

How does the "cover up" occur? Here Latour and Woolgar become very unclear; for, despite their lack of empathy with philosophy, they use a central philosophical notion in their account of the covering up process: the notion of a modality. In modal logic and metaphysics philosophers and logicians have worked out relatively precise characterizations of various modalities. Unfortunately, Latour and Woolgar are completely innocent of this prior work. They don't quite mean by 'modality' what philosophers and modal logicians mean by 'modality'. Just what they do mean is not entirely clear. A modality is a "mode" that qualifies the truth-conditions of a sentence. For example, 'possibly' is taken to set up a modal context. So, the sentence,

UFOs are spacecraft piloted by alien beings

does not mean the same (because it is not true or false under *exactly* the same conditions) as the sentence,

Possibly, UFOs are spacecraft piloted by alien beings.

Even if we knew that the first sentence is false, the second one could still be true—that's the difference that 'possibly' makes to the truth-conditions for what would be otherwise the same sentence. There are many kinds of nonmodal sentences that nonetheless function logically like modalities that the philosopher Bertrand Russell called **propositional attitudes**. Propositional attitudes have to do with the different psychological attitudes that a human being can take "toward" a sentence (actually, the attitude is taken toward the content of the sentence, which philosophers call the proposition of the sentence). For example, the sentence,

Some UFOs are spacecraft piloted by alien beings

does not mean the same as, indeed, it has very different truth-conditions, than the sentence,

Harvey is skeptical that some UFOs are spacecraft piloted by alien beings.

And both those sentences are yet different in meaning from the sentence,

Harvey is confident that some UFOs are spacecraft piloted by alien beings;

which is different again from the sentence,

Harvey wishes it weren't commonly believed that some UFOs are spacecraft piloted by alien beings.

I have put all the different propositional attitude indicator phrases in italics in the above examples. The reader can see the basic pattern. The same declarative sentence occurs after a different attitude phrase ending with the word 'that'. This is the commonly accepted way of representing propositional attitudes. Notice that the truth-value of the buried declarative sentence can stay the same while each sentence as a whole can change truth-value depending on the content of the attitude phrase. Suppose, for example, that as a matter of fact it is false that some UFOs are spacecraft piloted by alien beings. Nevertheless, it can be true that Harvey is skeptical of this fact, false that he is confident of that fact, and true that he wishes this falsehood weren't so commonly believed to be true by so many other people. The reader can now sense the virtually limitless number of different propositional attitudes: 'believes that ...', 'fears that ...', 'knows that ...', 'doubts that ...', and so on, are all propositional attitudes in which a declarative sentence fills the '...' position. Here is the important point about propositional attitudes for purposes of our present discussion: The truth or falsehood of a propositional attitude sentence is not purely a function of the truth or falsehood of the embedded declarative sentence—rather, the specific content of the attitude factors in also.

Latour and Woolgar are misled by the presence of propositional attitudes in scientific reasoning and practice into the belief that all scientific facts are about their own construction, not about some practice-independent reality. What misleads them is how different propositional attitudes occur at different times in the process of discovering a scientific fact. The attitudes expressed early in the discovery process tend to be attitudes involving skepticism and uncertainty. Surely this ought to be no surprise. Equally, it ought to be no surprise, that if scientists are fortunate and their conjectures are true, nature rewards their researches with success, and they begin to express attitudes toward the same "fact" that involve increasing confidence and sureness. For example, a research paper appearing in an immunology journal in 1987 might contain the following propositional attitude sentence,

We now suspect that NK cells kill some of their targets with the aid of antibodies.

This sentence indicates a tentative attitude toward the alleged fact in question. It is being surmised, floated as a possibility that subsequent practice will either support or reject. We can imagine it happening that by 1998 the same embedded sentence appears in a research paper attached to a more confident attitude,

We now believe beyond doubt that NK cells kill some of their targets with the aid of antibodies.

Latour and Woolgar take this evolution of attitudes from uncertainty to confidence to show that the belief scientists have that they are discovering preexisting material realities is a delusion; for, according to Latour and Woolgar, the move to more confident attitudes is a purely social creation, one having to do with how many times a given paper is cited in other papers, in textbooks, and in conversation, how effectively the prestige of one scientist can intimidate others into accepting his or her views, how standardized lab machinery becomes so that the same "wavy lines" are reproducible in different labs, and so on, and not one having to do with uncovering a fixed material reality. Their argument is instructively erroneous enough to be worth quoting at length:

> In simple terms, [scientists] were more convinced that an inscription unambiguously related to a substance "out there," if a similar inscription could also be found. In the same way, an important factor in the acceptance of a statement was the recognition by others of another statement that was sim-

> ilar. The combination of two or more apparently similar statements concretised the existence of some external object or objective condition of which statements were taken to be indicators. Sources of "subjectivity" thus disappeared in the face of more than one statement, and the initial statement could be taken at face value without qualification. . . . An "object" was thus achieved through the superimposition of several statements or documents in such a way that all the statements were seen to relate to something outside of, or beyond, the reader's or author's subjectivity (*Laboratory Life*, pp. 83–84).

External objects are said to be "concretised" inventions due to the repeatability of a given "statement" in the research literature. Objects are said to be "achieved" as constructions out of socially produced data that are repeatedly obtainable only because of the standardization of research equipment and procedures. It is no wonder then that later in their study Latour and Woolgar call the objectivity of the concretised, achieved material reality a mere "impression of objectivity" (*Laboratory Life*, p. 90). It isn't real objectivity at all, is the clear implication. The scientists are as deluded as the tribal witch doctors. If this isn't antirealism by the case method with a vengeance, one wonders what would be.

It is important for us to understand what has happened to the allegedly objective world in the hands of Latour and Woolgar. Here again, we see at work the pernicious influence of the Kantian distinction between phenomena and noumena. *The noumenal world is taken to be so epistemologically remote by Latour and Woolgar as to be never represented in the output of our detecting devices at any stage of our detecting procedures.* Put a slightly different way that makes clearer the antirealism, *the amount of "filtration" through artificial manipulation that scientific data undergo during collection is so great as to destroy whatever was represented about the objective world in the original input at the distal end of the data collection process.* We might call this "The Great Filtration" or "The Great Distortion." It has a distinctly Kantian flavor to it, and I have illustrated it in Figure 8.1. If one seriously believes in The Great Filtration/Distortion, then it will follow as a matter of course that science is about its own process of construction rather than about an objective world. Again, I quote Latour and Woolgar at length:

> Despite the fact that our scientists held the belief that the inscriptions could be representations or indicators of some entity with an independent existence "out there," we have argued that such entities were constituted solely through the use of these inscriptions. It is not simply that differences between curves [on superimposed graphs] indicate the presence of a substance; rather the substance is identical with perceived differences between curves. In order to stress this point, we have eschewed the use of expressions such as "the substance was discovered by using a bioassay" or "the object was found as a result of identifying differences between two peaks." To employ such expressions would be to convey the misleading impression that the presence of certain objects was a pregiven and that such objects merely awaited the timely revelation of their existence by scientists. By contrast, we do not conceive of scientists using various strategies as pulling back the curtain on

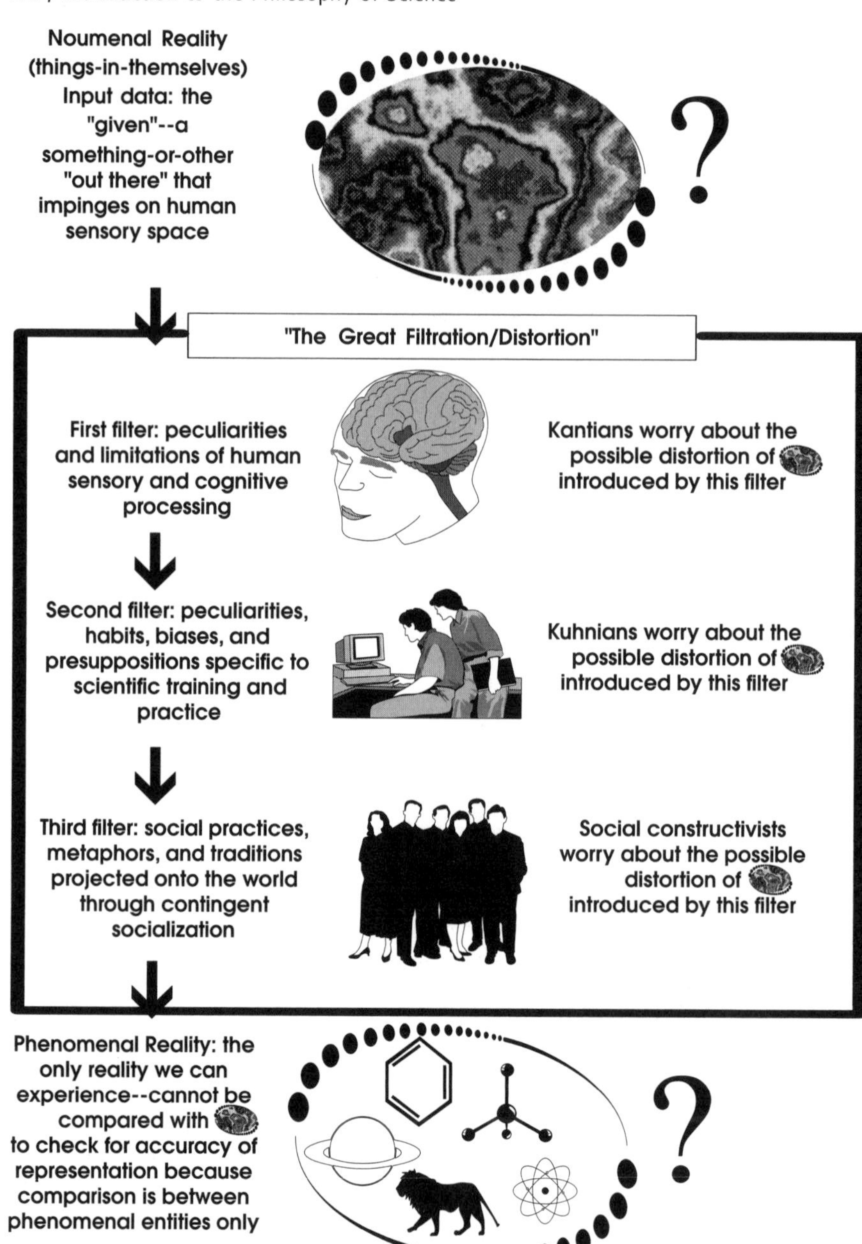

Figure 8.1. The Great Filtration/Distortion.

> pregiven, but hithereto concealed, truths. Rather, objects (in this case, substances) are constituted through the artful creativity of scientists (*Laboratory Life*, pp. 128–29).

One hardly knows what to make of the claim that a substance is "identical" with perceived differences between superimposed graphed curves. Bloor promised that the sociology of science would not be a form of idealism, but Latour and Woolgar in the above quote would seem to be wavering on the precipice of idealism. Substances are said to be identical with certain perceptual experiences. One doesn't know whether to laugh or cry when Latour and Woolgar tell the story of TRF (thyrotropin releasing factor, a neurohormone) as though TRF, the actual physical hormone itself (not merely facts *about* it), was created ("a major ontological change took place" *Laboratory Life*, p. 147); "a decisive metamorphosis occurred in the nature of the constructed object" *Laboratory Life*, p. 148) sometime between 1962 and 1969 in two particular laboratories.

Latour and Woolgar claim that the upshot of their particular brand of social constructivism is that *reality is the consequence of the content of science, not its cause.* Because they hold this view, they object to the idea that one could explain why different scientists arrive at similar or convergent beliefs by appealing to an independent world whose stable structure is the same for all scientists; rather, Latour and Woolgar insist, a common independent world is the entirely constructed outcome of the practices of scientists. It follows that realism gets it backward: The structure of the world doesn't constrain the content of science, the content of science constrains the structure of the world. The realist philosopher of science Philip Kitcher argues that for Latour and Woolgar to make good on this last claim they would have to show that differences in scientists' "encounters with the world" (read, with a noumenal, independent material reality) would make no difference to the outcome of what counts as the scientific facts, that no matter how those encounters with the world had been different, scientific facts would have been constructed with the same content. This, Kitcher argues, the two sociologists of knowledge fail to show with their case studies. I might add that what those case studies do show is that differences in the constructed content of scientific facts make a difference to the quality of our encounters with the world. But one could find enlightened realists who would agree with that. If Kitcher is correct, then Latour and Woolgar's interpretation of their study done at the Salk Institute is underargued and a rather misleading exaggeration.

In Latour and Woolgar's view, far from science being constrained in its content by a preexisting world "out there" whose structure it slowly discovers, science is a self-justifying collection of arcane technological practices—replete with mystifying jargon, guild-like exclusivity of participation, and all the trappings of ideological self-centeredness—a collection of arcane technological practices that literally invents the world from scratch. This is nuclear-strength antirealism if anything is.

Latour and Woolgar fess up to their antirealism eventually when they get around to a brief discussion of who they regard as the opposition. They identify the enemy of the sociology of knowledge to be what they call "a realist theory of science" (*Laboratory Life*, p. 178). Their argument against realism is so woolly as to be nearly unrecoverable, but their basic idea is that realism is a delusion. Their main reason for holding that realism is a delusion seems to be the old howler of a view that every philosophy teacher since the formal teaching of philosophy began has encountered coming out of the mouths of novice students in introductory-level philosophy courses: Namely, because you can't *talk about* an objective and independent material world without using concepts formed during the socially determined investigation of such a world, there is no such world at all—there are only the socially determined practices of investigation that produce the said concepts and the talk we make with them. The mistaken reasoning here is blatant and irreparable.

Latour and Woolgar's argument: If there were an independent material world, it would have to be directly knowable through nonlinguistic, nonconceptual means. But the second we start to talk about what we claim to know of such a world, we are helping ourselves to socially constructed concepts; and, if the concepts are socially constructed, so also are what they are concepts about.

The last clause in this argument is the sort of mistake that ought to remind the reader of one of Bishop Berkeley's more embarrassing moments: Because you can't think that trees exist independently of being thought about without, like, *thinking* about them, there are no such trees—there are only thoughts about trees; or rather, trees "just are" nothing but thoughts about trees. The way out from under this woolly argument of Latour and Woolgar's is surely pretty obvious: The first sentence of the argument is just plain false. Who says that if there is an independent material world it must be knowable directly by nonlinguistic, nonconceptual means? Certainly no realist would make such an implausible demand. Only the Kantian idealist would think up such a self-defeating and question-begging requirement. But Latour and Woolgar, if they are anything philosophical at all, are true to their Kantian (hence, to their idealist) roots: For them what exists collapses into what is knowable—in good Kantian idealist fashion—precisely because for them what exists collapses into what is investigatable, and what is investigatable collapses into what is describable, and what is describable collapses into what is knowable. In this tortuous way, Latour and Woolgar lead the sociology of knowledge up a dead-end alley into full-blooded antirealism, an antirealism that comes perilously close to what Bloor said the sociology of knowledge must avoid: an idealist model of science.

8.4 The View from Deep in the Swamp of History

It was inevitable that Kuhn's historicizing of philosophy of science would lead to a host of extremely detailed studies of particular episodes in the history of science. Book-length studies were surely destined to appear in which scholars would meticulously pour over the small-scale historical trivia of science: Galileo's private letters to his mistresses, Einstein's eating habits between thought-experiments, Darwin's sexual inhibitions as revealed by remarks in his journal on the breeding of domestic animals. I do not know whether any treatises have been published on the three above topics, but one social constructivist historical study did appear that caused a bit of a stir: Steven Shapin and Simon Schaffer's examination of the dispute between Robert Boyle and Thomas Hobbes over Boyle's air-pump and what it allegedly showed about the role of experimentation in science, a dispute that started in 1660 and lasted until Hobbes's death in 1679. The standard view of this dispute makes Boyle out to be the hero and Hobbes the villain. Boyle's position is taken to be the correct one while Hobbes's position is often portrayed as that of a boorish crank's, as though Hobbes was a mere armchair philosopher (whereas Boyle was a laboratory scientist) pathetically railing against the legitimacy of scientific experimentation from his moldy apriorist perch.

Shapin and Schaffer beg to differ. They seek in their study to show that Boyle didn't know as much as he has been thought to have known, that Hobbes was not quite the boorish crank the standard view says he was, and that Boyle's victory in the dispute was not due to any genuine superiority of his position vis-à-vis the objective facts about the physical world. Hobbes's rejection of Boyle's interpretations of certain experiments with his air-pump was cogent on its own terms, claim Shapin and Schaffer, and the wrongness of Hobbes's position is not much more than partisan history of science as written by the followers of Boyle. That is certainly an ambitious goal to aim their study at, and if Shapin and Schaffer could pull it off, then we should all be grateful for their having rectified a huge mistake in the heretofore accepted understanding of a critically important episode in the history of science.

In the midseventeenth century science was not yet understood as an empirical form of inquiry clearly distinguished from traditional philosophy. In fact it was common to refer to what we would now call science by the label "natural philosophy." As an example of how mingled together science and philosophy were, the philosopher Thomas Hobbes in his main work published in 1655 *defined* philosophy as the study of the causes of things. No philosopher nowadays would find that definition plausible at all: Philosophy is not a causal enterprise, it is an analytical form of inquiry, one that seeks to clarify, systematize, and criticize the concepts and methods of other more specialized domains of inquiry. The distinction we now take for granted between issues involving the conventional definitions of words and issues involving matters of fact about physical reality was not nearly as clear in 1660, even to the best minds of the times. It was a time when that distinction was first being worked out by human beings.

Robert Boyle was a brilliant natural philosopher who championed a new way of gathering knowledge, a way that did not fit the traditional form knowledge gathering took in established philosophy. Boyle was suspicious of grand a priori systems, of philosophers who presented complicated deductive systems of definitional truths that the structure of the world must supposedly match because reason itself supposedly necessitated such a match. A grand a priori system was precisely what Hobbes had labored to produce over many years and presented to the world in his monumental *De Corpore* in 1655. Against such overbearing systems of metaphysics Boyle advocated something brand new: the experimental matter of fact. Where the former kinds of metaphysical systems were all doubtful because they were entirely speculative, the latter facts of the matter were certain because they were directly witnessed through ordinary sense perception. Boyle made a distinction between knowing what a matter of fact is and knowing what its cause is. The former kind of knowledge is available to inquiring humans who perform the right sorts of experiments, the latter sort of knowledge is largely forbidden humans because it is beyond their godgiven means of perception to ascertain the hidden structures of things. Hobbes was of course aghast at such a distinction, for it meant that properly philosophical knowledge—knowledge of causes—was for the most part inaccessible to human beings. Hobbes was suspicious of experimentation, of mechanically inclined tinkerers who invented an artificial machine and then claimed to have discovered this and that about the world on the basis of the machine's behavior under artificial conditions.

Obviously, bad blood was bound to flow between Hobbes and Boyle. Boyle invented an air-pump, a machine he claimed evacuated almost all the air from an enclosed receptacle. He published a book reporting the results of forty-three experiments performed with the air-pump. This was a watershed moment in the history of science; for with Boyle's air-pump we see the arrival on the scene of two permanent institutional features of modern science: the standardized physical laboratory as a public space for experimentation, and the official skepticism toward all things theoretical (as opposed to all things observational) that became a universal requirement of the scientific mind-set—so universal and so required that Karl Popper later made it the very heart of scientific method with his falsificationism. Boyle's air-pump, as Shapin and Schaffer show, was the "big-money" science of its time, the seventeenth-century equivalent of a modern day subatomic particle accelerator or a space shuttle. So fancy was it, so difficult and expensive to design and produce was it, that no more than four or five working air-pumps ever existed in the entire world during the first decade of the Boyle/Hobbes dispute. By modern standards, of course, making Boyle's air-pump would be a very achievable project for a bright student competing in a school science fair. No matter, Boyle's air-pump was once the center of an intellectual firestorm during the birth pangs of modern science. Boyle claimed that some of his experiments with the air-pump proved certain incontestable "matters of fact": that the air has a "spring" to it, that the air has a weight to it different from its spring, that a vacuum is possible in nature, and so plenism (the philosophical position that a vacuum in nature is impossible on a priori grounds) is false. Hobbes was an "anti-vacuist," one of the armchair philosophers who held that a vacuum was impossible on grounds of pure reason. Naturally, Hobbes objected to Boyle's claim that the air-pump experiments involved the creation of a vacuum or near-vacuum inside the pumped receptacle. Shapin and Schaffer present the subsequent dispute in all its floridly bitter detail. A host of side characters appear in the lengthy drama, among them the Dutch natural philosopher Christian Huygens, Robert Hooke, a Jesuit priest named Franciscus Linus, and Henry More. Round and round they bickered and argued, experimented and counterexperimented (accept for Hobbes and his fellow philosopher Spinoza who both scoffed at experimentation as inferior to pure reason), and wrote treatises, tracts, and diatribes. Hobbes insisted that Boyle's air-pump leaked—indeed, that it *must* leak on a priori grounds—for a vacuum requires self-moving matter, which is a self-contradiction. Boyle insisted that the air-pump's leakage was negligible, that it created a vacuum for all *practical* purposes, and that the vacuum it created was a certain matter of fact on which all experimentalists could agree because it was publicly witnessable and repeatable. Nothing less than what form of life science would take was at stake, according to Shapin and Schaffer: Would it be the experimental form of life, or the form of life of a priori metaphysical system building? We can know only observational facts for sure, said Boyle, we are restricted to mere speculation about the underlying causes. Knowledge is of causes and how they fit into a grand metaphysical system or it is of nothing, said Hobbes, the so-called observational facts are mere artifacts of the artificial machines used to create them.

8.5 The Inequality of Theoretical Alternatives

Shapin and Schaffer's study is interesting and important. It shows just how conceptually messy science can be, especially when it was in its infancy and nothing could be taken for granted about it. The two historians take the underdetermination of theory by observational data as an established given throughout their analysis of the Boyle/Hobbes dispute. What is puzzling in their analysis is the lack of any recognition on their part that just because there exist alternative theoretical explanations for a given collection of observational data doesn't mean that all those alternative explanations are equally plausible and equally likely to be correct. Shapin and Schaffer seem to believe that there can be no principled way of ranking competing alternative accounts of observational data; therefore, they take it as established that *if* there are, say, six alternative and mutually incompatible theoretical explanations of a particular observational datum, there can be no question of one of those six theoretical explanations being any better than the other five. But why should anyone believe that? *Not all theoretical alternatives are equal.* Some alternative explanations cost too much in terms of changes we would have to make in other beliefs. Some alternative explanations postulate new objects or properties that we have no other reason to suppose exist, and so the postulation merely to explain this one observational data-set is *ad hoc.* Some alternative explanations increase the structural complexity of the system under investigation over what other alternative explanations do, and so the former explanations run a higher prior probability of being false. The ways in which competing alternative theories can be unequal with respect to their consequences for methodology, epistemology, and ontology are legion. Shapin and Schaffer seem unaware of this feature of competition between theories. They jump from the fact that Hobbes offered an alternative theoretical account of Boyle's air-pump experiments to the conclusion that Hobbes's account is just as plausible as Boyle's account. But there is no reason to suppose this kind of equality exists among competing theories. In fact, we know that it is extremely difficult to show that two theories are *exactly observationally equivalent.* The only known cases come from rare episodes in the history of physics. But even if two theories can be shown to be exactly observationally equivalent, it doesn't follow from such equivalence alone that the two theories are equivalent with respect to their methodological, epistemological, and ontological implications. We saw in chapter 4 how Quine, who is very fond of the underdetermination of theory by observational data, nevertheless presents a host of principles that can be used to rank competing theories: simplicity, conservatism, modesty, and fertility. Arguments can be given for why theories that maximize simplicity, consistency with our established beliefs (conservatism), modesty, and fertility at generating novel predictions are more likely to be self-correcting toward the truth over the long run than theories that do not maximize those pragmatic virtues. Hobbes's alternative account of the air-pump experiments did not win out over its rivals in the end. In fact, no theories that postulated a subtle fluid aether—of which Hobbes's was a variety—survived into the twentieth century. They had too many other epistemological costs that would have made it unwise to keep them afloat. Shapin and Schaffer conveniently leave out this subsequent history of atmospheric science. The suggestion at the heart of their study is that if there had been a few key differences in the sociopolitical factors that made Boyle's view triumph, then we would be Hobbesians this very day as you read this—that Hobbes's theories failed not because they were false but because the contingent politics of early science went against them. Scientific truth is a function of human sociopolitical intrigue much more than it is a function of what the objective structure of the world is like, is Shapin and Schaffer's point. As they say near the end of their study, it is "not reality which is responsible for what we know" (*Leviathan and the Air-Pump*, p. 344).

Once again, the principle of the underdetermination of theory by observational data is being asked to support a form of antirealism that it does not logically entail. This argumentative maneuver is a standard strategy of social constructivists in every discipline. Their assumption seems to be that if nature itself does not unambiguously point to a unique theoretical explanation for something, then there can be no principled way of claiming that there can be an objective truth about that something. But this conditional is unfounded. We can now say with supremely high confidence that:

1. Boyle was correct that the air has weight.
2. Boyle was correct that vacuums are possible.
3. Boyle was incorrect that the air has "spring."

4. Boyle was incorrect that the air contains a subtle aether.
5. Hobbes was incorrect in certain claims about the air (it contains a subtle aether, vacuums are impossible, and all chemical phenomena are mechanical in nature).

Shapin and Schaffer would apparently have us believe that these truths are neither objective nor apolitical; and Latour and Woolgar would add that the above truths only *seem* to be objective and independent of sociopolitical determination because the official mythology of science covers up their socially constructed pedigree. It is neither true nor false, per se, that the air has a spring, according to the social constructivists—at best, it is false *relative to* the sociopolitical framework of our current scientific form of life. But, had that sociopolitical framework and its attendant form of life been different in certain ways, then it would be true that the air has a spring. This last claim might strike others besides myself as more than just a little fantastic. But if it were true, then the science critic is potentially greatly empowered to change society through changing its dominant form of knowledge gathering: science. If the facts—*all* the facts—can be made different by making the social conditions of their construction different, then it is a short step for those social constructivists who harbor an activist streak to see their critiques of science as a form of revolutionary social activism. That is one reason why there is a certain kind of science critic who doesn't mind a model of science in which the input of an independent material world has been reduced to insignificance or eliminated altogether, who is in fact very glad to have arrived at that result. For, to such a science critic, the end-product of scientific endeavor under such a model of science is entirely or almost entirely a function of complicated social behavior; and, given that the content (or almost all the content) of science is therefore determined socially, that content is potentially alterable by altering the social conditions of its production. Thus, the possibility arises of altering the substantive content of science through political activism that seeks to alter the social conditions under which the production of scientific knowledge takes place. Some social constructivists, such as Bloor, are not specifically interested in this further activist aspect of the sociology of knowledge. They seem satisfied to point out the social determination processes by which the content of science is produced. We might describe this by saying that those social constructivists stress the *social* in social constructivism. Other social constructivists, in particular the feminist science critics of the next chapter, stress the *constructivism* aspect of social constructivism, the view that the factual content of science can to a considerable extent be constrained by self-conscious social engineering, by deliberate intervention in the scientific process for purposes of furthering a partisan political agenda. What I want to suggest here is how the nonfeminist social constructivist accounts of science naturally lead, via openly idealist variants of the sociology of knowledge, to the more radically interventionist feminist models of science. The feminist science critics who are responsible for those more interventionist models of science are the topic of the next chapter.

Further Readings

Social constructivist studies of science are legion, but the three discussed in this chapter are among the more interesting ones. David Bloor's Strong Programme is perhaps the clearest and best thought-through social constructivist model of science. It is set out in his *Knowledge and Social Imagery*, 2nd. ed., Chicago, University of Chicago Press, 1991. Bruno Latour and Steve Woolgar's account of their time at the Salk Institute is related in *Laboratory Life: The Construction of Scientific Facts*, 2nd. ed., Princeton, Princeton University Press, 1986. A very sophisticated social constructivist view of the Boyle/Hobbes dispute is in Steven Shapin and Simon Schaffer, *Leviathan and the Air-Pump: Hobbes, Boyle, and the Experimental Life*, Princeton, Princeton University Press, 1985.

9

The Politics of Epistemology

9.1 The Genderization of Scientific Knowledge

Social constructivism as a philosophy of science suggests that it is not wise to give too much credence to how those who participate in science understand it from the inside. Indeed, social constructivism advises us to be especially skeptical and wary of "insider accounts" of science. All such accounts are suspected of being self-serving and lacking the proper insight into the sociopolitical nature of science and its intellectual products. If the content of scientific theories is not a function of the objective facts, but rather of the socially constructed methods of daily practice, then science's claim to give us the universe "as it really is" is so much exaggeration. What it gives us is what the socially constructed methods of daily practice license as a *claim* about what the universe is like; and the funny thing about socially constructed methods of daily practice is that they are contingent historical accidents, rules of method that are not read off of nature itself as much as they are manufactured piecemeal by the fallible descendants of great apes that we are. Once this point is reached, it then seems important to consider precisely who gets to do the social constructing of these methods of practice. Such a consideration, to some science critics, soon produces a rather controversial conclusion: science looks suspiciously "cooked up." To these critics it begins to look like science is simply one more power-wielding social institution controlled by the privileged rulers of modern, technology-based social life to further their own interests. And, who are these privileged rulers? There can be only one answer: practitioners of science steeped in the traditions of male-sexist European culture, for patriarchal European culture was the birthplace of science in the form in which it has come down to us in the present day. Science criticism from the radical regions of the social sciences and humanities thus ends up joining forces with a formidable companion: contemporary postmodernist feminism. Perhaps nothing at the present moment provides a better case study of this sort of criticism of science than does the work of the philosopher of science Sandra Harding. Let us take a closer look at her model of science as our entree to what its proponents call **feminist science criticism**.

9.2 Postmodernist Feminism: Sandra Harding

Like any product manufactured by the social labor of imperfect organisms, science has a checkered track record. That science has had some awfully ugly and stupid moments ought to be neither a surprise nor something necessarily worthy in every case of eternal shame. The same checkered track records can be observed in the histories of any of our other hallowed social institutions, from religion to athletics to art. Yes indeed, a mere century ago mainstream science considered masturbation to be a serious mental disease, criminality to be a causal result of skull shape, and intelligence to be unevenly distributed by race. The list can be expanded to cover everything from the absurd (Paracelsus: the element mercury cures syphilis because the planet Mercury is the astrological sign of the marketplace and syphilis is contracted in the marketplace where the prostitutes who carry it are found) to the sublime (Freud: clinical paranoia is caused by repressed homosexuality) in the ways of error-making.

It is the *kinds* of errors that science has made that the feminist science critics are concerned to highlight. These kinds of errors, so their account goes, are not randomly distributed; rather, they cluster around certain predictable elements in what is held to be a uniquely male mind-set. Science as done by men bears the distortions of male psychoneurosis. The methods of daily practice by which the game of science proceeds bear the structural psychological faults and foibles of their creators—men. The result, according to the feminist Sandra Harding, is a scientific methodology and worldview in which the psychological defense mechanisms and alienated affections of men are read back into the phenomena of nature as though they were microstructural realities woven into the very fabric of things instead of the merely neurotic projections that they really are. The projections are performed mostly unconsciously, and they ultimately serve to maintain the set of de facto social power relations in the contemporary world, social power relations which keep males in charge of most of the important institutions and resources of our planet.

Harding allies her portrait of science with what she calls postmodernist feminism. One identifying feature of postmodernism is its insistence on denying the legitimacy of various dichotomies that it traces to the European Enlightenment:

object/subject, fact/value, mind/body, culture/nature, and so on. These pairs of concepts form the structural basis of the male European worldview and mind-set, the same worldview and mind-set that invented and still controls the methods of daily practice by which science proceeds, the same methods of daily practice by whose use the facts of nature are supposedly discovered to be what they are independently of what anyone thinks, wants, or wills them to be. The prevailing male "wisdom" is that one of the two concepts in each pair is to be preferred to the other member of that pair. Science is then to be designed with respect to its methodology so that it maximizes the preferred member of each pair. Feminism adds the observation that it is no accident which member of each pair gets elevated to the preferred status: It will be that member of the pair culturally associated under patriarchal forms of thought with maleness, while the remaining member of the pair, the nonpreferred member, is associated with femaleness. Men are *objective*, care about the *facts*, as those facts are ascertainable intellectually by the rational *mind*, in such a way as to constitute intellectual *culture*. Women, according to this patriarchal mind-set, are *subjective*, care about the *values* of facts, as those values relate to the life of the material *body*, as that material life of the body flows out of *nature* itself. The distinguishing of men from women in terms of intellectual traits is rampant throughout the feminist literature, and Harding cites such literature heavily. Women are said to be more intuitive than men, they think qualitatively where men think quantitatively, and they think concretely where men are obsessed with abstractions. These various differences are linked to supposedly universal early life experiences in a neo-Freudian way (Harding is fond of a small splinter group of neo-Freudian theorists called object-relations theorists). Because sexist child-rearing practices entail that women do a disproportionately larger amount of child-rearing than men, little boys are at greater psychological risk than little girls when it comes time to separate from the main parental care giving figure and form their own identities (by contrast, little girls do not need to disidentify first with the mother then reidentify with the father). Little boys mostly fail in this project and develop the classic male personality pattern as a neurotic defense: that is, issues of autonomy and individuality dominate their thinking, rather than issues of connection and belonging, which organize the thinking of females. In Harding's hands, it is a short step from this universal psychological damage to a world of male-made science in which the universe is conceived to be a heartless and cold spatial void filled with individual atoms swirling around without purpose except when they are being buffeted about by exclusively external forces. Apparently—on Freudian grounds—modern elementary particle physics is, for the most part, externalized, institutionalized, male sickness of soul. There is throughout Harding's work a significant hostility to physical science, as contrasted with social science, and the hostility is based mostly on the heavier use physical science makes of mathematics and other systems of abstract concepts. Harding denigrates physical science on the grounds that it is excessively, even *perversely*, abstract, and that it consists of male-invented, "mystifying" abstractions that mean little to women—or at least they mean little to the kinds of feminist women who have not been intellectually co-opted by the male rulers of modern science. This is an increasingly popular charge against physical science in certain quarters, and the reader is referred to Figure 9.1 for an illustration of it.

In short, men are basically too psychologically damaged to be the ones doing most of the science. It would be better if women did most of the science. In what sense of 'better', we might ask? In the sense of being more objective is Harding's startling reply. She is being a tad ironic here. As an article of feminist faith she holds that women are feelers of the concretely subjective, not reasoners about the abstractly objective—like men—so she cannot mean by 'objective' what men mean by 'objective' (for thinkers who supposedly reject the popular conceptual dichotomies, it is striking how frequently feminists presuppose those very dichotomies they officially reject). Indeed, she argues that to be objective in a properly feminist way means to be subjective in a politically progressive way. Objectivity in science is increased, *not* by value-neutrality, but by allying science

1. Albert Einstein's gravitational field equation is,

$$R_{ij}-\tfrac{1}{2}g_{ij}R=8\pi GT_{ij}.$$

This equation relates the metric tensor g_{ij} to the stress-energy tensor T_{ij}. Oversimplifying greatly, the metric tensor is a representation of the geometric structure of spacetime (in particular, distance intervals between events), the stress-energy tensor is a representation of the distribution of matter in spacetime, G is Newton's gravitational constant (6.67 x 10^{-11} Newtons), R_{ij} is the Ricci tensor and R is the scalar Riemannian curvature, which each represent aspects of the curvature of spacetime. The metric tensor alone contains 16 variables.

2. The Friedman equation for the expansion of the universe from its initial "big-bang" beginning is,

$$H^2-\frac{8\pi}{3}Gd=-kR^{-2}.$$

This equation has something to do with the "fate" of the universe, with whether it will expand forever or eventually contract, or neither. H is the Hubble parameter, which measures the rate of expansion of the universe (currently estimated to be about 20 kilometers per second per 1,000,000 light-years), G is again Newton's gravitational constant, d is the average density of matter in the universe (currently estimated to be between 10^{-31} grams per cubic centimeter and 5 x 10^{-31} grams per cubic centimeter), R is a scale factor that measures the amount of expansion so far, and k is a constant that represents the curvature of spacetime and is either 1, -1, or 0.

Figure 9.1. An example of the allegedly perverse abstraction of male-made physics.

openly and forcefully with liberation movements throughout society and the globe. At this point we encounter an infusion of Marxist social theory. Just as it was the only socioeconomic class without a vested interest in maintaining the status quo—the proletariat—that alone, according to Marxist theory, was capable of leading the political revolution to a socialist state, so it is the groups and classes currently shut out from science in the contemporary world that are alone, according to Harding, capable of leading science to a better and more accurate portrait of the universe. The idea seems to be this: The holding of certain political and moral beliefs on *other* issues, beliefs whose contents derive from having suffered marginalization and oppression, is guaranteed to bring with it greater accuracy in purely scientific beliefs. Specifically, all errors in science (or at least a large majority of them) are due to the vestiges of racism, sexism, classism, and homophobia that surreptitiously infect science at its very foundations. These evils infect science at its very foundations because science is the product of the European male mind, and Harding takes it for granted that a majority of European males were and are racists, sexists, classists, and homophobes. A science cleansed of all traces of European male racism, male sexism, male classism, and male homophobia, is a more accurate science, factually speaking. This claim is unargued for by Harding except insofar as she takes the historical cases in which those four moral evils produced stupidly inaccurate science as showing that their eradication will produce accurate science. This form of argument is of course the fallacy of denying the antecedent. From that fact that 'All A is B', it does *not* follow that 'All nonA is nonB'. So, from the fact that all science produced by the moral evils mentioned is inaccurate science, it does not follow that all science produced in their absence (and replacement by their opposites) will be 'not inaccurate', that is, accurate. Perhaps Harding's response would be to reject the logical principle that denying the antecedent is fallacious; after all, what is logic itself but one more product of the male mind invented to serve its interests in maintaining hegemony? The last thing Harding will be impressed with is a claim that logic is to be paid attention to because it is the paradigm case of a value-neutral methodology. From her perspective, "value-neutrality" in science is a reactionary concept used by the threatened male power elite to belittle and declare bogus the alternative perspectives on the universe held by the underclasses of the earth.

Harding recommends a strong tonic to correct these gender-based distortions in science, one that cures at the same time as it demonstrates the diagnosis. Because value-neutrality is a myth that psychologically damaged male neurotics need in order to invent their science, it would make sense that they would enshrine the least value-laden science as the King of Sciences—mathematical physics—while at the same time they would relegate the most value-laden of the sciences—the social sciences—to second-class status. Harding recommends a complete reversal of the classical Unity of Science program that the positivists advocated. Actually, all the sciences *are* hierarchically related, suggests Harding, but in the reverse order from what the positivists thought: mathematical physics is not the most fundamental domain of inquiry, the domain to which everything else must be reduced; rather, the social sciences are the most fundamental domain of inquiry, and everything else shall be put into their service. Value reigns supreme over fact. The moral and the emotional suffuse all things that dwell in and on the earth and in the heavens above—even, we must presume, a seemingly inanimate hydrogen atom out in interstellar space (but that's a male abstraction anyway, so why should it matter if the new shoe doesn't quite fit that case?).

What are we to make of this account of science? Suppose, for the sake of argument, that Harding's basic claims were accurate. Suppose her account of science is really how science is despite the popular fairy tales about it to the contrary, despite the alternative models of it to the contrary, and despite what many scientists themselves insist it is. The reader ought to notice something interesting in that case: There would seem to be little in the way of an external world pushing back against our theorizing, an external world whose structural nature is independent of our pet moral beliefs. One of the functions the concept of "the factual" (as opposed to "the value-laden") was invented to perform was to represent the externality of the nonhuman part of the world, to represent the truth that we don't always get to make up how things are, to represent the *fact* that the world is not any one person's oyster, that it does not obey one's every whim and command, and that it often resists our attempts to manipulate it for the betterment of human life. Now, of course, here Harding might be tempted to launch into an **ad hominem** counterattack on me: "Why do I focus right away on aggressive manipulation? Why do I immediately seize upon the language of command as though I need to dominate rather than get along with nature? These are such typically male fetishes and obsessions." Harding and a number of other feminist science critics seem a bit too cavalier about the scandalously high frequency of such *ad hominem* personal attacks in their own technical writings. There is a kind of jaunty haughtiness that infests the feminist science criticism literature and which encourages the proliferation of such *ad hominems* with a perilous disregard for the negative cost their use exacts: to wit, argument by *ad hominem* switches the focus from the original topic and the entire discourse veers off target. Perhaps a good and nasty lick or two can be gotten in on the opposition as persons, but their argument remains unaddressed and is still on the table. In the present case, that argument culminated in my charge that an independent nature has all but disappeared under the postmodernist feminist critique, that there is no external world left "out there," that reality becomes something almost totally plastic, something we can make into almost anything we wish it to be. In short, the original topic was my argument to the conclusion that postmodernist feminist science criticism is necessarily committed to a form of extreme antirealism in the philosophy of science, that it is committed to the view that elementary particle physics, to take one of Harding's favorite targets, might simply be entirely fictitious. There don't *have* to be atoms and subatomic particles and weak nuclear forces and strong nuclear forces, and what have you, Harding argues, for these terms do not refer to real pieces of external material structure in the universe, rather they refer to gender-specific, metaphorical concepts projected onto nature to meet sick male psychological needs. In general, Harding considers realism to be simply an equally neurotic "successor project" to the old and thoroughly discredited positivist model of science, and her attitude toward realism is accordingly hostile. Notice also how the notion of evidence seems to drop off the map. Harding expresses little interest in whether an overwhelmingly huge weight of solid experimental evidence might support these basic scientific concepts she so enthusiastically attempts to consign to the scrap heap of gender-biased ideology. "Evidence is relative," will be her response, "it's all cooked up, anyway—it's all a tight question-begging circle"; or, as Harding suggests in an analogy to the computer programmer's saying "garbage in, garbage out" (what errors the computer makes are only the ones you wrote into the program it executes in the first place): male-made science is a case of gender distortion in, gender distortion out. This broad-scale, across-the-board antirealist conclusion to her analysis of science seems to be something that either goes unnoticed by Harding, or else it is rather cavalierly embraced in passing as though it did not have the evidentially implausible and politically Orwellian implications it does.

Some of Harding's critics have pointed to what they see as a lack of substantial case-study evidence in her work, a disturbing sparseness of case specifics, which forces her to assume an ever more extreme and purely ideological bent. Some of that may be due to her commitment to Marxist and neo-Freudian modes of analysis, areas in which it has long been notorious to find a relative dearth of well-conceived empirical studies and a relative glut of free-form speculative theorizing. One pair of critics, the scientists Paul Gross and Norman Levitt, recently accused Harding of an essential ignorance of the basic methods and substantive facts of physical science. Their argument seems to be that she counts as a case of the old cliche that what a person doesn't understand she tends to fear and loathe. It is hard to tell if this is a fair charge against Harding, for it is hard to tell from her work whether she really is ignorant of physical science or whether her silence on its specifics is a voluntary choice made in support of her preference for the social sciences. Perhaps she could quote Newton and Einstein chapter and verse, as they say, but she chooses not to out of a superior regard for the more "womanly" sciences. (Actually, I suspect that she has had some sort of exposure to Newtonian physics, for at one point she suggests that Newton's laws of physics might really best be understood as constituting a "rape manual" [*The Science Question in Feminism*, p. 113]. On the theory that you have to know something about what you are misrepresenting to *mis*represent it, this remark would argue for a prior familiarity with Newtonian physics.)

Another recent critic of feminist epistemology, Cassandra Pinnick, has argued that Harding and her sister feminist science critics present no empirical evidence for their centrally important claim that when marginalized and oppressed persons do science the substantive results are more "objective" than when the male power elite does science. The charge here seems to be that this is an empirical question, yet feminist science critics like Harding fail to present the necessary empirical studies which could back up their preferred answer. We will return to this criticism later.

From my point of view, the most philosophically implausible thing about Harding's analysis of science is what I have called its extreme antirealism, the cavalierly embraced implication that doing accurate science is a question of putting politically progressive people in charge of making up the facts and the methods that will produce them (and the reader should recall here how at the end of chapter 7 I argued that it is not obviously the case that Kuhn's model of science necessarily commits a Kuhnian to this kind of blanket scientific antirealism—more argument is required to establish it than simply citing Kuhn's work). Harding's position ought to remind the reader of what we saw the nonfeminist social constructivists argue in the previous chapter. If David Bloor is correct, then truth never acts as a constraint to explain why specific scientific claims become accepted. If the truth of those beliefs doesn't explain why they are held, then the door is opened for Harding to argue that their popularity is due to factors that are not evidence driven at all, but driven by irrational biases and prejudices, the biases and prejudices of the gender that currently dominates science—men. If Latour and Woolgar are correct, then the "objects" that science constructs are literally "concretised" via sociopolitical negotiations among practitioners. Harding adds the gender of those negotiators into the mix, and the result is her claim that the facts are malleable by sociopolitical means. Hence, it makes sense for Harding to argue that the content of science is ours to make, and therefore that the issue that really matters in the end is who this 'our' denotes. Science is disguised politics, seems to be Harding's ultimate claim.

It is a sign of our times that science now gets charged with being a surrogate form of politics. The charge several decades ago was that science was disguised religion. What one senses is that the former, currently fashionable, charge is as much a caricature of science as the latter more traditional charge. Science is not something else other than science, and it needs to be understood for itself and in itself. Harding's model of science does not facilitate such a genuine understanding of science. Her portrait of science turns it into something the rest of us can hardly recognize.

All is not lost for feminist science criticism, however. Feminism is no more monolithic a movement than any other large-scale social revolution, and there are as many different feminist approaches to the philosophy of science as there are different feminists. Harding herself acknowledges that the feminist science criticism literature suffers from a kind of conceptual woolliness, a seemingly permanent inexactitude and inconsistency of style, rhetoric, and substantive doctrine. She locates the source of this conceptual woolliness in feminism at large—

in its general canon of literature—and she suggests that it is a natural result of feminism's postmodernist structure. Postmodernist feminism, as she understands it, denies that there is an essential female type, a standard female character that is the same in all cultures and contexts (it remains both a mystery and an illustration of Harding's very point about inconsistency in feminist thought how one could square this denial of a characteristic female personality type with the common feminist practice of attributing certain standardized cognitive traits to females—qualitative thinking, concrete thinking, greater emotionality, and so on). Men wish to believe the contrary—that all women are the same—for it serves their neurotic need to objectify women and turn them into alien "others" who can then be dominated without guilt. Harding claims that in reality women suffer from "fractured identities" (and men presumably do not?—but we have been told how psychologically damaged they all are!), identities that are culled, cobbled, and jury-rigged together out of the cross-solidarities which the experience of patriarchal oppression forces resisting women to forge with others. She calls these feminism's "hyphens," and she asserts that they are to be defended at all costs. There are no just plain feminists, there are only lesbian-feminists, African-feminists, muslim-feminists, battered-wife-feminists, rape-victim-feminists, working-class-feminists, bisexual-feminists, socialist-feminists, academic-feminists, and so on. Each kind of feminism generates its own view of things; and the natural result is a canon of feminist literature notoriously full of emotionally rich but conceptually woolly and highly perspectival claims. Accordingly, maximum consistency and uniformity of view are not the virtues at which feminist literature aims. Harding promises her readers that this is a good thing about the feminist canon of literature. I propose to take Harding at her word on this matter and to look closely at the work of two other feminist philosophers of science, with the hope that their views will be less ideologically implausible than Harding's and more congenial to the commonsense realism of ordinary scientific practice.

9.3 Feminist Social Activism: Helen Longino

Helen Longino associates her feminist model of science with the holist camp in the philosophy of science. Holists have cause to be wary, however. Longino argues that holism implies that evidential relations are entirely disconnected from causal relations, that is, that what counts as confirmatory evidence or disconfirmatory evidence for what else is entirely independent of and unrelated to what states of affairs in the world are causally connected to what other states of affairs in the world. Longino's claim is that what causes what depends on what else you already believe, on your background beliefs. But the holistic model of science implies no such disconnection between causal relations and confirmation relations. What holism implies is that evidential relations are a posteriori (that is, knowable only after inquiry proceeds, not knowable prior to the beginning of inquiry), not that they are totally disconnected from causal relations. Nothing could be more misleading than to suggest that Quine's holistic model of scientific theories gives no role to causal relations in the confirmation or disconfirmation of scientific hypotheses. On the contrary, what fixes evidential relations for a Quinean holist could be nothing other than the de facto causal relations that happen to obtain as brute empirical regularities among the phenomena. But because she denies the coherence of the very notion of a belief-independent causal reality, Longino openly embraces what she thinks are the relativist and antirealist implications of holism.

Like all consistent relativists and antirealists Longino is cut off from talking about objectively existing determinative connections and regularities and restricted to speaking only of *what we believe to be* objectively existing determinative connections and regularities. Longino argues that we are trapped within our world of socially derived beliefs—learning about objective reality is an impossible goal, for we cannot break out of the shell of socially produced beliefs. The object of scientific inquiry is never nature, Longino argues, but "nature under some description." What interests Longino is how those descriptions of nature become distorted by institutionalized forms of bias that flow from ensconced but unjust social traditions and practices. Our current science is distorted in specific ways, she argues, because the larger society in which science functions does not constitute a just social order. Science would look different, have different methods and results, under more ideal conditions of social justice than those that obtain right now. Surely no one familiar with the checkered track record of science could disagree with her on this point, taken as a matter of the history of science. The controversial part of Longino's analysis of science is her further claim that the distortion is much more widespread and insidious than popularly believed, that it has not substantially decreased over time, and that only a feminism-informed mode of criticism can lead science out of its current level of distortion and closer to how science was meant to function in a truly just social order.

Longino presents two main case illustrations of how specifically male gender-biases have produced inaccurate science. Her first case is from anthropology. There is a dispute among practitioners in that domain about the causal origin of tool-making and tool-using among our hominid ancestors. One theory, the "androcentric" theory, claims that the development of tool-making among intelligent apes was due to "man-the-hunter," who needed lethal weapons to slay the savage beasts of the African savannah. A second theory, the "gynecentric" theory, claims that tool-making developed among intelligent apes because of "woman-the-gatherer," who needed tools to scrounge and forage for scarce vegetarian food. Obviously, the truth lies in some mean between these extremist theories. That someone would adamantly insist on an exclusively sex-differentiated origin for as general a phenomenon as tool-making and tool-using would seem to smack of clinical obsessionality. Longino's discussion of this dispute seems to take it far too seriously, as though she really thought that a well-trained anthropologist could seriously think that the woman-the-gatherer theory was the truth, the *whole* truth, and *nothing but* the truth. It is hard to see how that position would be any improvement over the man-the-hunter theory, because it would simply be the converse extreme, taken to the extreme. What Longino's overserious discussion inadvertently exposes is the overt political-interestedness behind the feminist critiques of the man-the-hunter theory. What really matters to those feminist critics is not what the fossil evidence suggests (Longino conveniently provides an argument that any evidence can be made to support any hypothesis you want it to—just change enough background beliefs), but rather what really matters is ushering in a properly feminist anthropology in which traditional male aggressiveness among our ancestors is not glorified and institutionalized in the theories taught as the truth in anthropology textbooks. A person does not have to be a defender of the man-the hunter account to see the questionableness of rushing to support some equally oversimple countertheory that replaces the glorification of male aggressiveness with the glorification of a "gentler, kinder" female as an anthropological type.

The second case illustration Longino presents is from biological science, and it involves the investigation of how mammalian prenatal hormones affect the subsequent development of the fetus, particularly with respect to alleged cognitive differences between males and females. Longino charges the scientists who did most of the experimental work on prenatal hormones with bringing gender distortion into their otherwise allegedly pure experimental investigation of hormonal influences on cognitive differences between the sexes. One complaint seems to amount to the familiar charge about the dangers of extrapolating experimental data from nonhuman animals—in this case rodents—to humans. Longino's complaint here is a bit undermotivated. Rodents, she says, don't have a behavioral repertoire as rich as humans. After all, humans dwell in a social milieu suffused with intentionality, rodents presumably do not. One could hardly disagree with this. Rodents don't hate calculus, admire the current governor of the state, or know how to make baklava, yet many humans beings are capable of performing one or more of those intentional actions. But how does that show, even probatively, that rodent neurophysiology is sufficiently distinct from human neurophysiology as to make hormonal experiments on rodents *irrelevant* to the behavioral effects of hormones in human beings, particularly if we are talking about the *same* hormones, biochemically speaking? Indeed, we can talk about the same hormones, biochemically speaking, precisely because of the significant degree of phylogenetic similarity between rodents and humans that follows from their both being mammals. We did have a common ancestor—we humans and rodents—but you'd never think so reading Longino's remarks about the rodent studies in question. Such a complaint in the end betrays a failure to appreciate facts about common phylogenetic descent, comparative anatomy across species, and the fundamentals of mammalian neurochemistry.

Longino's main target is prenatal hormone research on human subjects. Such studies usually involve a population of humans who suffer from some form of congenital hormonal dysfunction. For example, females born with a condition known as CAH, congenital adrenocortical hyperplasia, are exposed in the womb to excessive amounts of hormones that are classified as androgens. Follow-up studies seem to suggest that such prenatal hormonal imbalances produce down-stream *behavioral* effects (there are clear anatomical effects in severe cases, which no one denies). CAH females are, to use the admittedly questionable term thrown around in the research literature on CAH, "tomboyish" from early childhood on. CAH females are biochemically inclined to behave like tomboys, is the common view in this area, and Longino is rightfully offended by the loaded language of it all. Her critique is interesting for the way in which it highlights her anti-realism. She points out that the variables of research observation in this case are already gender-laden. 'Tomboy' is not a value-neutral term. What is going on is that the researchers are already adopting the hypothesis under test in the matter of CAH females. The researchers already believe that the prenatal hormonal imbalance produces sex-differentiated behavior, and so, charges Longino, they impose a **taxonomy** (a scheme of classification) of observable properties on the behavior of these females that is guaranteed to highlight what they expect to find. Her suggestion is that a different taxonomy would not cough up the same regularities, the same tomboyish behaviors, as the one used in the original research. On the contrary, she argues, researchers only found tomboyish behavior because the concepts in terms of which they observed the behavior of CAH females presupposed tomboyish behavior. If they had looked for something different, seems to be her point, then they would have found something different. This last point is important. Longino accepts without apparent argument the proposition that there is no nonpolitical, nonmoral way to rank these different taxonomies of observational properties with respect to their epistemological accuracy, and hence that there is no correct fact of the matter, as that notion is traditionally conceived, to be discovered about whether prenatal excesses of androgens produce certain childhood and adult behaviors in females. Notice how Longino differs here from Harding. On Harding's view, feminist taxonomies are clearly superior to all other taxonomies on grounds of epistemological accuracy, and feminists have access to substantive general truths denied all nonfeminists; so, for Harding, there is almost always a properly feminist fact of the matter to be known about something, a view to which Longino does not openly commit herself. Longino's attitude, not explicitly written but clearly implied by what she does write, is that there are no "facts" in the traditional sense of that concept, at least not in this area involving prenatal hormones and behavior. There are only claims that are socially constructed, politically defeasible, and manufacturable through the use of the proper taxonomy. Once again, any sign of an external world of relatively stable structural realities that pushes back against our theorizing has withered away, and we are left with only the theorizing, with only a supposedly highly plastic system of adjustable concepts and classifications that we can juggle to suit our social and political purposes free of any causal constraints imposed by a nonhuman nature. This ought to remind the reader of Shapin and Schaffer's mistake that we reviewed in the previous chapter. Shapin and Schaffer assumed that all alternative explanations for a given phenomenon are equal with respect to their epistemological, methodological, and ontological aspects. Longino makes the same error with respect to alternative taxonomies of observation: She appears to believe that there is no way of ranking alternative taxonomies with respect to their different epistemological, methodological, or ontological virtues (for they are all equal in those respects). The only way to rank them, if we're to believe Longino, is by their different *political and moral* implications.

Longino describes her own perspective in the area of biologically based cognitive differences between the sexes in valorous terms, terms designed to indicate that *her* perspective is not simply that, another subjective perspective that, for all we know, might harbor its own kind of truth-destroying distortion. Longino argues that the substantive results of science must be constructed so as to "valorize"—her chosen word—the subjective experiences of oppressed persons. In Longino's view we must have a brand of science that allows for maximal self-determination and maximal self-validation of the individual, *no matter what*. If the best evidence points to the contrary conclusion—to the view that some of us are defective, dysfunctional, or just plain nonperfectable—we will simply reinterpret that evidence. If the best available theory points to the personally demoralizing conclusion, we will dump it for a different theory that gets us the morally comfortable result we wish to see enshrined as scientific fact. The substantive results of science, according to Longino, should make us feel good about ourselves. As Longino puts it bluntly while discussing what is wrong with male-made brain science,

> Our [Longino's and her colleague Ruth Doell's] political commitments, however, presuppose a certain understanding of human action, so that when faced with a conflict between these commitments and a particular model of brain-behavior relationships we allow the political commitments to guide the choice. (*Science as Social Knowledge*, pp. 190–91)

In all my years of practicing philosophy of science I have rarely run into a more candidly millenarian personal statement of relativist and antirealist sympathies than when Longino writes,

> The idea of a value-free science presupposes that the object of inquiry is given in and by nature, whereas contextual analysis shows that such objects are constituted in part by social needs and interests that become encoded in the assumptions of research programs. Instead of remaining passive with respect to the data and what the data suggest, we can, therefore, acknowledge our ability to affect the course of knowledge and fashion or favor research programs that are consistent with the values we express in the rest of our lives. From this perspective the idea of a value-free science is not just empty but pernicious. (*Science as Social Knowledge*, p. 191)

Notice that Longino is careful to say that it is not merely our theories *of* the objects of inquiry that are partly socially constructed—you could find positivists who'd agree to that—but rather it is those very objects our theories are theories of that are partly socially constructed. This is precisely the same error we saw Latour and Woolgar commit in the previous chapter. According to social constructivism, there is no principled distinction between our theory of X's and X's themselves. But this mistake seems no more reasonable in a feminist context than it did before in a nonfeminist one. How could the Andromeda galaxy be socially constructed (even partially) when it is over 2 million light-years (over 12×10^{18} miles) distant from the earth? How could it be that the internal microstructure of a mammalian NK cell *is ours to choose* (even partially) so that it might be compatible with our moral and political needs? The manifest implausibility of both suggestions reveals the falsehood of the same conditional premise implicit in both Latour and Woolgar's argument and in Longino's argument. That our theories are socially constructed does not by itself entail that the objects our theories are about are socially constructed, even partially so. This shared error between the nonfeminist social constructivists and the feminist social constructivists provides an explanation of why it is that Harding and Longino join the nonfeminist social constructivists in being quite fond of Kuhnian/holistic models of science; for it is such models that, under their interpretation of them, allow for the possibility of always recooking the evidence so as to make it evidence for a different kind of object of inquiry—if you don't like the kind of object of inquiry the old evidence suggested, that is.

The feminist science critic is invested with this kind of potential "world-making" power in the interest of furthering politically progressive culture through politically progressive science. Longino does not trust the traditional self-correction mechanisms in standard science—they cannot be counted on to churn out politically progressive results, in her view. The social constructivism inherent in this lack of trust is fairly clear. The world, according to the social constructivist view of it, is not a "pregiven" (independently structured) cosmos largely indifferent to our human desires and moral aspirations; rather, from Longino's viewpoint, the world is a morality-friendly and partially malleable place where we are at home (or at least, it ought to be such a place under a politically enlightened view of it). Philosophers of science who are not social constructivists can agree wholeheartedly with Longino on her political vision and moral aspirations while also arguing that there is no reason why that vision and those aspirations require that science be mobilized and seemingly co-opted so as to guarantee the reinforcement of our social ideals. For one thing, if those social ideals are justifiable social ideals (which most of them surely are), then they need no extra reinforcement from the empirical universe—they stand on their own intrinsic merits. There is no need to reverse the error of making morality a hostage to science by making science a hostage to morality. For this reason, Longino's attempt to co-opt science for laudatory moral purposes is misguided because it distorts the independent legitimacy of morality and creates a hostage out of science where no such hostage is needed. Just as you can't legislate morality, neither can you discover it empirically. In this regard, Longino's laudable ethical motives have led her slightly off the track about the nature of science. Yes, Longino is on our side. Yes, let the moral truth be shouted out from the mountain top. Yes, as Chairman Mao said it so well, let a hundred flowers bloom. Make love, not war. But, even though it would make the world a wonderful place if some of the most rationally defensible moral views an enlightened person can hold were true, their truth would seem to be something having virtually nothing to do with what's up in the vast nonmoral parts of nature: with whether the Hubble parameter (the current rate of expansion of the universe per unit distance) is 20 kilometers per second per 1 million light-years or twice that amount, with whether NK cells kill cancer cells solely by perforin puncture or whether they have a second way they can do it, with whether the hole in the Antarctic ozone layer has tripled in size since last year or only doubled in size, or with whether the symptoms of acute schizophrenia causally supervene on heightened sensitivity among an average number of dopamine receptors in the brain, an increased number of dopamine receptors of average sensitivity in the brain, or both, or neither.

The point I am making seems not to be completely unfamiliar to Longino. In some of her work she tries to adopt what reads like an old-fashioned positivist distinction between the raw observational data and our theoretical interpretations of it, even though her supposed holistic proclivities would suggest that such a distinction is bogus. She argues, for example, that her model of science does not imply that whether the needle of a gauge on a fancy scientific device points to 10 is up to practitioners to say, that her model is consistent with the claim that whether or not the needle points to 10 is not primarily a function of the socially constructed background beliefs practitioners bring to the observation of the gauge. I think that she is certainly correct about the objectivity of where gauge needles point, but it directly contradicts her remark quoted above that the object of scientific inquiry is not given in and by nature. For objects are composed of properties, and what this hypothetical gauge is detecting is one such property. Longino's eleventh-hour attempt here to convert to a positivist epistemology is rather questionable. If whether or not the needle on the gauge points to 10 is not primarily a function of our socially constructed practices, theories, and the background beliefs they license, then it would have to be primarily a function of certain objective structures in nature; then it must be that the needle is attached to a device that detects an objective property of the system being measured, an objective property given in and by nature, not given in and by our communitarian habits of inquiry and practice. But that is just the sort of situation she had ruled out in her remark previously mentioned; that is, for the object of inquiry not to be given in and by nature is just for its properties (or, at a minimum, its essential properties) not to be given in and by nature. Longino even tries in one place to claim for her model of science a consistency with what she calls a "minimalist form of realism." The resulting form of realism would surely have to be minimalist to the extreme. One doesn't know quite what to make of Longino's claim here, but it certainly does separate her from Harding and the nonfeminist social constructivists, all of whom reject realism emphatically. I must confess that I do not see how realism of any kind can be squared with the rest of Longino's model of science, but I remain hopeful that I am wrong and that someday she will make it clear and understandable to all of us how it can be so squared. In the meantime, we must suspect that she is engaging in a bit of philosophical backpedaling.

What one senses may be the source of Longino's backpedaling is her struggle with the same preoccupation that flavors Harding's work: scientific topics that directly bear on issues of sexuality. All her main examples come from such areas of research. There is in their work a heavy focus on case illustrations having to do with research in psychosexuality, homosexuality, transsexuality, sociosexuality, gender disorders, and the whole sexy lot. Of course these are perfectly wonderful and important research topics in their own right—but they are not the whole of science, not even the mainstream of science in its current form. That Longino and Harding would both overfocus so much on this one area of research, that they would both overgeneralize to the whole of science from errors made in this one area of inquiry, smacks of private-agenda analysis; and it does so in the same way, I might suggest, that the positivist model of science smacks of private-agenda distortion when it obsessively overfocuses on the physical sciences. From my point of view, the reader who seeks a realistic model of science instead of highbrow political ideology would do well not to buy what Longino and Harding are selling: a heavily ideological portrait of science, one that does not inspire confidence that we are getting an evenhanded view of science. But I still hold out hope that feminist science criticism can be put into a form more congenial with science's overall success at what it does than the forms Harding and Longino offer us. This hope of mine may not be without just cause; for feminism comes in many varieties, and we may find a more plausible feminist portrait of science in the work of Lynn Hankinson Nelson.

9.4 Feminist Empiricism: Lynn Hankinson Nelson

Lynn Hankinson Nelson identifies herself as a Quinean, an adherent of the philosophical views of W. V. O. Quine, whose influential philosophy of science we investigated in chapter 4. She also identifies her model of science as a form of empiricism, albeit a feminist empiricism. That this is a brave thing for Nelson to do flows from certain remarks Sandra Harding and other feminist science critics have made suggesting that 'empiricism' ought to be something of a dirty word for feminists. In Harding's view, so-called feminist empiricism is the weakest kind of feminist science criticism; for it stays too close to the established, gender-distorted mainstream in philosophy of science (recall that Harding adopts postmodernist feminist science criticism as her personal approach).

Nevertheless, we ought to be solidly confident about Nelson; for Quine was absolutely batty about science, and it is hard to imagine how any Quinean could be hostile to science and still be a Quinean. Indeed, the early going in Nelson's model of science is encouraging. She is openly concerned—and what a relief it is to see her so concerned, in contrast to Harding and Longino—that an external world of causally fixed structure that pushes back against our theorizing not be banished from the philosophy of science in the name of political and social empowerment of women, or for some other morally laudable ulterior motive. What Nelson wants to do, instead, is to undermine a distinction that Longino and Harding both take for granted and then both choose the same side of to defend: the distinction between discovering the facts of science and socially constructing them. Where Harding and Longino each opt for defending the social construction of scientific knowledge over its discovery, Nelson argues that knowledge gathering in science proceeds through a combination of both social construction and discovery—that, in a sense to be rendered consistent with a Quinean form of realism, the distinction itself is obsolete in the philosophy of science. It is worthwhile to dwell on the caveat. Many people assume that a Quinean holist must be some kind of antirealist. This is odd, for Quine has made it abundantly clear that he is not a scientific antirealist (yet the contrary seems to be the consensus view among non-Quineans). As we saw in chapter 4, that a theoretical posit is performed on pragmatic grounds is not sufficient to conclude that it is the posit of an "unreal" entity. What distinguishes the Quinean realist from the more traditional sort of scientific realist is that the Quinean holds that fallible pragmatic principles of method, if properly pursued long enough, are capable of providing plausible grounds for holding that our mature scientific theories give us a picture of "the actual way things are," whereas the traditional view holds such principles of method to be incapable of such a result no matter how properly or for how long they are pursued. The traditional realist aspires to a system of scientific method that provides a way of getting outside the web of belief to compare reality directly with what is claimed about it from inside the web of belief. Quine denies *both* that we can get outside the web of belief to make such a comparison and that this fact cuts us off from being realists—such was his genius to have concocted a philosophy of science that seeks to make consistent both claims.

Nelson analyses what she argues are two mistaken presuppositions of the positivist and neoempiricist traditions in philosophy of science. Their first mistake is their commitment to an observational/nonobservational distinction, a distinction the reader gathers from Nelson's arguments that is ultimately without merit. This is encouraging, again, not only because it is a plausible claim, but because other *realists* (such as Grover Maxwell) had said the same thing previously; that is, if there is one thing an *anti*realist must adopt to make antirealism tenable it is a clear distinction between the observational and the nonobservational (see section 10.3.2 on the antirealist Bas van Fraassen). So, here is some probative evidence that Nelson is drifting toward a realist model of science. The second mistake of the two traditions is their assumption that philosophy of science can proceed fruitfully by ignoring the so-called "context of discovery" as contrasted with the so-called "context of justification." I have chosen not to emphasize this standard distinction in this text for a number of well-chosen reasons, many of which Nelson illustrates in the course of her discussion of it. She argues that how scientific ideas are discovered and arrived at is not as irrelevant to their subsequent testing and justification as the positivist and neoempiricist traditions assumed. It is a main feature of Nelson's view that *who* does science is extremely critical, and that one central failure in the traditional models of science is their refusal to recognize that fact. This is an interesting claim, for one would have thought—as, indeed, the positivists did think—that the subsequent justification procedures in science—the "back end" of the scientific process—would act as a correcting brake on most biases introduced at the discovery end of the process of knowledge gathering. Nelson denies that the means of justifying and testing scientific ideas and claims are sufficient by themselves to correct biases loaded in at the "front end" of the scientific process. This case must be clearly made out, and Nelson's presentation of her model of science aims to do just that.

Nelson interprets the work of Kuhn differently than I did in chapter 7. In my view she underappreciates the stabilizing effects of paradigms: how it is that, because paradigms settle fundamentals, normal science is thereby opened up for extremely precise research into minute details. The sheer volume and complexity of the details produces the plodding sameness of normal science. This, I argued, was the real source of normal science's inherent conservativeness and rigidity of practice. Nelson seems unaware of this source and attributes the conservative rigidity of normal science to what sounds like peer pressure—as if it is due to sociological factors like conforming to the way the powerful peer group does things, fear of opposing one's elders, habituation of professional training, and so forth. This seems a mite cruel toward practitioners, for the majority of them are not nearly as personally cowed by their peers or their previous education as Nelson's view suggests (on the contrary, science may be overpopulated with especially egomaniacal people). Nelson reads Kuhn as a rank antirealist and relativist, foisting on him the very strongest interpretation of incommensurability she can and implying in no uncertain terms that Kuhn is committed to a model of science under which it is in the end an entirely sociological enterprise, divorced from the objective world and insulated from the larger society in which it takes place. I am not sure this is fair to Kuhn (although his own flip-flops and fence-sitting in *The Structure of Scientific Revolutions* are mostly to blame for the fact that such a strong reading can be given of his work), but that doesn't matter really because Nelson is no fan of Kuhnian philosophy of science (again, unlike Harding and Longino). She rejects both the positivist tradition in philosophy of science *and* the Kuhnian countertradition embraced by nearly all of her sister feminist science critics. This certainly stakes out for Nelson a unique location on the feminist playing field.

Despite her Quinean proclivities Nelson identifies with the sense of epistemological self-importance that is a frequent trademark of contemporary feminist science criticism. It is startling to find someone who purports to be a Quinean adopting Harding's view (actually due to the work of the feminist Nancy Hartsock and ultimately of Marxist origin) that feminists, simply in virtue of being feminists, have cognitive access to substantive general truths denied all nonfeminists, that they have what Nelson calls "epistemological privileges." She tries to soften the implausibility of this kind of special pleading by stressing that the privileged knowledge in question is not due to unbridgeable epistemological chasms between feminists and nonfeminists. No, Nelson argues, everyone has the potential to become a feminist science critic, if only they want to bad enough and try hard enough—even men. The point seems out of place in what purports to be a commonsense empiricist model of science.

On the other hand, Nelson does not believe that relativism in either epistemology or ontology is a viable option. She is concerned that her sister feminist science critics not cave in to relativism, that they not be intellectually hoodwinked by the sociologists of knowledge into embracing it; for relativism, she argues, simply cannot be rendered plausible with respect to the overall success of science. This is as close to an argument for scientific realism that Nelson gets in her model, and I suggest that we take it as such. It bears certain structural similarities to one of the standard arguments for scientific realism that we will investigate in chapter 10. Hence, I will assume from now on that she is some sort of realist. Accordingly, Nelson sees that in the end it is *truth* at which science aims. Harding seems basically uninterested in truth, as indeed any genuine postmodernist would be, and Longino restructures issues of truth as issues of political justice and empowerment.

Nelson puts forward the clearest body of evidence for making the case against male-made science of any of the three feminist science critics surveyed in this chapter. Nelson cites four illustrative examples: (i) the methodology in traditional social science of taking male specimens to represent an entire species; (ii) the work of the feminist psychologist Carol Gilligan on differences in moral and ethical thinking between females and males; (iii) the massive literature on research into biologically based differences in cognitive abilities between the sexes; and (iv) the man-the-hunter/woman-the gatherer dispute in anthropology. The majority of Nelson's criticisms are applied against (iii). She presents evidence that the research into biologically based differences in cognitive abilities between the sexes has turned up null results: There are no such differences. So be it. There obviously was no way to know this prior to sustained inquiry into the matter. But Nelson writes as if such research should never have been undertaken in the first place. This suggestion is oddly out of place in a Quinean model of science; for the implication of Nelson's complaint is that *if* differences had been found, they would have been unacceptable with respect to favored political and social doctrines (and therefore not true?). Nelson argues that the null results could have been anticipated without the research and that the fact the research went on anyway shows how psychologically disreputable the males who undertook it were and are. Those males were hoping to find such differences, is Nelson's insinuation, so that such differences could be used to support de facto inequalities between the sexes presently ensconced in social life. The idea is that it is a specifically male desire to read contingent social inequities into nature, to find such relationships "natural" in some biological sense. The whole discussion reads like a version of the old nature/nurture dispute, although Nelson insists that feminist theory denies the legitimacy of that traditional distinction. What ought to bother other Quineans is Nelson's attempt to stake out an area of inquiry—biologically based cognitive differences between the sexes—as somehow demarcating an area of nonempirical facts we can know about in advance, and for which empirical inquiry is therefore useless. Such an exception is nowhere called for under Quinean holism, and one wonders if Nelson's anger at the sexist research produced in this area temporarily sidetracked her from the Quinean path.

Nelson argues that feminist science criticism is part of science itself, that feminist science critics are *doing science* when they write their critiques. This licenses an interesting argument:

> All feminist science criticism is science,
>
> Some feminist science criticism is inherently political and moral,

therefore

> Some science is inherently political and moral.

Nelson is unabashedly optimistic about this conclusion, for she takes one implication of Quinean holism to be that science cannot be isolated from the political and moral realm of experience. In response to the objection that she can't possibly be an empiricist and hold such a view, she argues that political and moral inquiry, like any kind of inquiry under a Quinean model of holism, is under empirical constraints—it is not an "anything goes" area of inquiry. Unfortunately, her attempt to illustrate how moral and political critique are under empirical constraints comes close to the sort of feminist special pleading that is sprinkled throughout the work of Longino and Harding. The "commonsense experience" of women throughout history, argues Nelson, serves as an empirical test of the moral and political elements in science, an empirical constraint that corrects any distortion in that content in the direction of greater truth. As a general *method* one wonders how that is any different from a man claiming that the "commonsense experience" of men throughout history serves as an empirical constraint on the moral and political content of science, one that corrects any distortions in that content in the direction of greater truth. One does not correct distorting biases by introducing their opposites.

Quinean holism goes astray, Nelson argues, in taking the individual cognizer, the individual practitioner, as the unit of epistemological action. This is wrong not only in fact, she claims, but it is wrong politically; for such an individualistic empiricism in philosophy of science would be unable to accommodate the insights of specifically *feminist* science criticism, and those insights must be accommodated by any philosophy of science which aspires to even minimal adequacy. This is tied into the feminists-know-things-nonfeminists-don't business in the following way: It is in virtue of their uniquely feminism-informed *social* experience that feminists come to know things nonfeminists don't, so the social nature of that unique experience is clearly an important element in what underwrites its claim to provide a superior knowledge of things. As she puts it in so many words, the group identity of an individual is an important epistemological variable in evaluating the beliefs of that individual. Hence, we must "socialize" the Quinean model of epistemology. It is the community of practitioners—indeed, the entire society in which the practitioners are members—that participates in a group construction of our knowledge, and therefore of our scientific knowledge. *Individuals do not know anything of a specifically scientific nature*, only communities composed of individual practitioners can be said to know something as a piece of scientific knowledge. We can illustrate her point with an example from immunology. *I* can be said to know my name, my present address, and other such nonscientific items of belief; but *I*, all by myself, cannot be said to know that one of the main ways NK cells kill cancer cells is by perforin puncture. Only the larger community of immunologists working in partnership with other specialized departments of knowledge seeking can arrive at such a belief, which the rest of us can then share in having, teaching, communicating, and refining further. Nelson pushes for this communitarian view of epistemology under pressure supplied by Harding. Harding and the feminist Alison Jaggar have both taken aim at what they call "empiricism," and particularly "feminist empiricism." Nelson argues that antiempiricist feminist science critics like Harding and Jaggar can do this only because they buy into a certain assumption that derives from Enlightenment empiricism: that value-laden beliefs that arise from radical political activism on the part of the cognizer are not constrained by empirical evidence, that "anything goes" when it comes to moral and political beliefs formed by radical activism. That is not true, argues Nelson. There are objective constraints suggests Nelson, but they are not objective in the sense in which Enlightenment empiricism understood objectivity. *Who* it is that knows does matter, but who it is that knows turns out not to be any one or more individuals—not Sally or George, or Ahmad or Juanita—but the community as a whole. Individualistic Enlightenment empiricism, which saw knowledge gathering in science as a function of the summable efforts of many separate individuals, could factor out political and moral practices and beliefs from influencing the substantive content of science one individual at a time, as it were. But this cannot be done if the entity that gathers knowledge is the community as a whole, for a community's moral and political assumptions, practices, and beliefs are structurally intrinsic to everything it is and everything it does. Hence, a community's moral and political practices and beliefs, far from being irrelevant to the substantive content of science, percolate and radiate throughout the whole of science, affecting all its substantive content. Nevertheless, Nelson insists, there is an external world that pushes back against those communitarian practices, beliefs, and assumptions, so science is not be free to order up just any old thing as a scientific fact:

> We assume that there is a world, including a social world, that constrains what it is reasonable for us to believe because that assumption makes the most sense of what we experience. To be sure, we construct theories rather than discover them, and these theories are underdetermined by all the evidence we have for them. But it is not up to us whether gravity is real or whether many research programs are androcentric and this has led to distorted accounts of how things are. Beliefs and theories are shaped and constrained by public theory and their consistency with each other. (*Who Knows: From Quine to a Feminist Empiricism*, p. 295).

But, is that enough of a constraint in the end? Nelson does worry about the relativist and antirealist implications of the feminist science criticism program in general. In her view, she has saved Quinean feminist empiricism from relativism and antirealism by the move to a communitarian conception of epistemology. Only if it is individual cognizers who know, rather than communities, is there a problem reconciling empiricism with something she takes for granted must be made consistent with it: the fact that feminist science critics know things about the distortions of traditional science that nonfeminists don't. As she puts it,

> But, again, though insisting on the role played by community membership in the ability of feminists to 'know things others have not', I am not advocating a sociology of knowledge. I am advocating that we look for the relationship between what feminists know and their experiences within a real social order, and I am arguing that those experiences and a real social order constrain what it is *reasonable* to believe. (*Who Knows: From Quine to a Feminist Empiricism*, p. 297).

Harding argues that empiricist epistemology cannot be rendered compatible with the special knowledge, the "epistemologically privileged" status, possessed by feminists because, as Harding understands empiricism, it is isolated individuals who know things, and feminist knowledge is essentially communitarian knowledge: It flows from the uniquely feminism-informed social activism essential to feminist experience. Nelson answers, no problem, it is not individuals who know under a properly Quinean form of empiricist epistemology, so empiricism is not after all inconsistent with the epistemologically privileged consciousness of feminists. But now notice that no one has filled in the obvious missing piece of the picture: Why should we believe that feminists, *as feminists*, know things nonfeminists don't? It is no answer to be told, because the unique social character of feminist activist experience informs feminists that they are correct in their views, for that makes it sound like what is going on is that the traditional cliche about "women's special insight" is being dressed up in fancy philosophical party clothes. We feminist women "just know these things" hardly counts as an equitable or defensible epistemological argument. It is no more equitable or defensible than when Cardinal Bellarmine told Galileo that he "just knows" the Scriptures are correct when they claim that the earth does not move; nor is it more equitable or defensible than when Aristotle argued that any right-thinking person "just knows" that in all composite entities there is a ruling part and a subject part (and so slavery is natural). It is this very claim that feminist social experience produces greater objectivity and accuracy in knowledge-gathering pursuits ("epistemological privileges") that the feminist critic of feminist epistemology, Cassandra Pinnick, argues is the main flaw in feminist science criticism. This claim, according to Pinnick, surely must be an empirical one (it obviously is not a purely conceptual claim), and as such it would need to be supported by sound empirical studies that show the greater accuracy of feminism-informed scientific theorizing. But no such studies have been conducted by anyone, Pinnick points out, and so the claim hangs in the air as a pure conjecture, as activist rhetoric dressed up as philosophical insight.

If the gap between the claim to epistemological privileges and evidence for it could be filled in with substantive empirical support, then feminist science criticism, as it is currently practiced, would be a stronger and more formidable force in the philosophy of science. Nelson argues that her version of feminist science criticism is better situated to welcome the undertaking of such empirical studies into feminist epistemological privileges because her model of science makes values partly constitutive of empirical science itself. Values are not occupiers of a separate realm of truths (distinct from facts of the matter) unconstrained by empirical experience, rather they are of a piece with all the rest of our knowledge and subject to experience-based criticism. Hume would certainly turn over in his grave at all this; but, in a sense, he along with Descartes are the two main philosophical villains in the eyes of Nelson. The is/ought distinction is perhaps the real battlefield on which Nelson's feminist empiricism wars with the larger sexist culture in which it resides. It is no accident that Nelson finds herself remarking specifically on that familiar distinction,

> In large part what has sustained [the traditional model of science] has been a basic skepticism about values and politics, that they cannot be subjected to evaluation, and the view that science is an autonomous undertaking. Feminist criticism indicates we must reject both views. The suggestion that values can be objective is not new. An enormous amount of ink has been spilled in the past attempting to show that one or another set of a priori values embodies the objective truth about morality or politics. Many of us have found these uniformly unconvincing perhaps because, with a few notable exceptions, we have not taken seriously the view that knowledge is of a piece, that how things are in part determines how things ought to be, and that how things ought to be in part determines how we take things to be. It is time we did so. (*Who Knows: From Quine to a Feminist Empiricism*, pp. 316–17).

There is an interesting asymmetry in this epiphanous passage, which represents the superiority of Nelson's model of science over competing feminist models of science. Notice the second to last sentence in the passage. There Nelson suggests a certain symmetry between what is and what ought to be—or, does she? She gives the impression that she will reach for the symmetry but then sidesteps it at the last second. If the claim was to be perfectly symmetrical she *would* have written, ". . . that how things are in part determines how things ought to be, and that how things ought to be in part determines *how things are*" (my italics). But she changes the part I have italicized to, ". . . how we take things to be." Now, "how we take things to be" is not the same concept as "how things are." The gap between the two is wide in metaphysical and philosophical importance. Which is it, how we take things to be or how things are? Harding would reject the question outright on postmodernist grounds. Longino would pick the second option because she mistakenly thinks that adopting holism in epistemology means that we inquirers after nature can remake nature into what we'd like it to be. Nelson wisely chooses the first option, for those are her very words. Nelson refuses to play the game of feminism-informed "world-making" that her sister feminist science critics seem quite sympathetic to; for they are all social constructivists and antirealists whereas Nelson is a Quinean realist.

Nelson's model of science, while far from perfect, seems to this author to stake out the direction in which feminist science criticism would do best to go. I have indicated points at which I would take issue with Nelson; but her commonsense realism and her appreciation that not all competing theories are epistemologically, methodologically, or ontologically equal separate her, in a positive way, from the nonfeminist social constructivists of the previous chapter, while her commitment to empiricism separates her, even more positively, from almost all her sister feminist science critics, including Harding and Longino. Ultimately, time will rule on Nelson's optimism about there being truth-enhancing empirical constraints on our moral and political beliefs. If she is correct, then we may anticipate with relief and pleasure the development of an experience-based mode of moral/scientific life which is self-correcting toward greater truth despite the fact that the knowledge-gathering methods that mode of life employs are openly politically interested. There are many other thinkers unconverted to feminist science criticism who remain more skeptical and apprehensive about such an outcome. Part of that skepticism and apprehension grows out of the conviction that the events of the world roll on without regard to humanly ordained political values, that the world's structure is what it is largely independently of those values, that there is a preexisting reality that, to speak metaphorically, doesn't "care" about our pet moral and social conventions. This is a contentious conviction, one that in recent years has received much attention within philosophy of science. Some philosophers deny that the conviction can be sustained in the end. To this clash between the realists and the antirealists in the philosophy of science we turn in the next and final chapter of this book.

Further Readings

The natural starting point at which to enter the feminist science criticism literature is Sandra Harding's, *The Science Question in Feminism*, Ithaca, Cornell University Press, 1986. Helen Longino's model of science is found in *Science as Social Knowledge*, Princeton, Princeton University Press, 1990. Lynn Hankinson Nelson presents a feminist empiricist model of science in *Who Knows: From Quine to a Feminist Empiricism*, Philadelphia, Temple University Press, 1990. Cassandra Pinnick criticizes feminist science criticism in "Feminist Epistemology: Implications for Philosophy of Science," *Philosophy of Science 61* (1994): 646–57. A general critique of feminist science criticism that is interesting but perhaps a bit underargued is found in P. Gross and N. Levitt, *Higher Superstition: The Academic Left and Its Quarrels with Science*, Baltimore, The Johns Hopkins University Press, 1994, pp. 107–148.

10

The Actual Way Things Really Are

Science was born in the dank bogs of supernatural religion. The ancient Greek sages mixed empirical hypotheses with sometimes jarring supernatural assertions to produce the combination of guesswork and fantasy that passed for scientific wisdom in the ancient Greek world. As we saw in chapter 5, Thales is credited with being the first scientist, for having said that "the first principle of all things is water." But he is also said to have claimed that "all things are full of gods." Does this mean that the gods are watery? Well, who knows? His written works are lost to us, and we can only piece together from other sources a crude picture of Thales and what he supposedly said and did. For example, some philosophers have argued that the statement about all things being full of gods ought not to be interpreted in the religious sense we might otherwise be inclined to assign to it; that in fact Thales was being ironical, for he really meant to indicate that one does not need to postulate personlike beings (that is, "gods") inside ordinary things to explain their behavior. He was arguing *against* animistic modes of explanation such as those found in most religions by suggesting that the processes that explain things are intrinsic to them and not supplied by their being inhabited by godlike spirits. On this view of the remark, what Thales was saying was something like, "a crazy person is no more inhabited by a god than is an inanimate rock, and an inanimate rock is no less inhabited by a god than is a crazy person—so forget the appeal to gods to explain things, instead seek the explanation of things in terms of purely natural processes that are structurally intrinsic and nonpersonal." Other philosophers find this interpretation too modern minded, and Thales was hardly a modern-minded person. It is not legitimate to foist on him the sensibilities of a secularized Westerner of our own times, which is what the first interpretation does. According to this second view, Thales' scientific ideas could not be divorced from the background of traditional superstition that permeated his culture. If the latter view is correct, then so be it. What else could be expected of bright thinkers working in the dark, as it were, without the benefit of experimentation and advanced technological instrumentation? They did what any one of the rest of us would have done in their position: They filled in the huge gaps in what they knew about nature with speculation and, at times, superstition.

The result was, shall we say, less than sterling in quality. A slew of unobservable objects, properties, and beings were postulated by the ancient scientists, which proved to be a motley of mostly fantastic conceptions concocted by brilliant thinkers working in the nearly complete absence of empirical constraints on their speculation. Spirits, entelechies, souls, forms, substantial forms, the *apeiron* (the boundless), the homoeomers (don't even ask), microscopic polygonal figures, and even plain old numbers were among the many "fundamental building blocks" postulated as the structural/causal basis of things. In some cases we can barely understand these basic unobservables as consistent notions of any kind. The lack of success at positing unobservable structures that science suffered from at its beginning ought to be expected when there was relatively minuscule feedback from experiment or observation. Inquiry about the structure of the world thus got off to a sloppy start, and it has been a boot-strapping operation ever since—two steps forward and one step backward. This sort of track record for science naturally gives rise to a healthy and appropriate skepticism about the claims of science, especially about the ultimate microstructural story that science seeks to unravel. How will we ever know if we get it right? That is the question that eventually emerges as *the* question in the epistemology of science.

10.1 What Can or Cannot Happen in the Limit of Inquiry

One response to the question is to challenge the question's very coherence. Perhaps such a question presupposes certain controversial views about reality, views which themselves need to be examined and first agreed to before we need to worry about some question that presupposes them. The nineteenth-century philosopher Charles Sanders Peirce took just such an attitude toward the question. He is one of the founders of the philosophical movement called pragmatism, often said to be the only school of philosophy born in America. A pragmatist, among other things, is a philosopher who has very little patience with "pure concepts," very little tolerance for conceptual distinctions that don't "make a difference" to what anyone could actually experience in their sensory life. A pragmatist wants to know the experiential cash value, so to speak, of a conceptual distinction before admitting that distinction as legitimate. One of the many targets that Peirce attacked was the idea of a reality which outruns all the actual and possible experience human beings could ever have, a reality still said to be "there," but *so* hidden away, *so* remote, that it is forever cut off from our experience. Peirce thought that such a conception of the world was pointless and useless. Hence, he had little patience with otherworldly thinkers who postulated a reality beyond the human senses, a reality in which certain states of affairs were nevertheless true, were nevertheless one definite way rather than another way. To Peirce such truths would make no difference to anything we could experience, so they might as well not exist at all. Hence, he recharacterized the concepts of truth and reality along pragmatist lines as follows: What is true, what counts as reality, is whatever view of the world is fated to be the account of it upon which everyone would agree *in the limit of empirical inquiry*. Read for 'empirical' in the last clause 'scientific'. This notion of a limit of inquiry was a bit strange at the time when Peirce first used it, but recently the notion has come in handy in representing the substantive issue in contention between scientific realists and scientific antirealists.

Peirce's idea is really quite an elegant thought-experiment. Here is how it works. Let serious scientific inquiry proceed indefinitely into the future (never mind the worry that we might blow up the planet tomorrow—this is a thought-experiment). Such inquiry, being *scientific* inquiry, isn't just any old kind of inquiry. It's not like tea-leaf reading or crystal-ball gazing, after all. It is a self-correcting, communitarian, cautious, and technologically sophisticated form of inquiry that gets better and better over time in terms of its epistemological accuracy. Now this definite advance toward greater accuracy will never reach perfect accuracy, for inquiry may never come to an end and reach a final result. Here we can use the mathematical concept of a limit, an approach to a point that gets closer and closer to it indefinitely without ever actually reaching it (what is called an asymptotic approach). In this hypothetical limit of scientific inquiry, said Peirce, there will be some final form in which our theories are put; that final form, what our theories assert to be the case in the limit of inquiry, simply *is* what is true of reality. In other words, in the limit of inquiry, ontology collapses into communitarian epistemology: What everyone (that is, careful empirical inquirers using sound scientific method) would agree on in the limit of inquiry is what is true. There is no sense to imagining the possibility that our theories in the limit don't match some theory-independent structural reality. Peirce's view is a position that we saw earlier has come to be labeled antirealism. Antirealism is the view that the structure of the world is ultimately not independent of the structure of our theories of the world. The opposing position is called realism, the view that, even in the limit of inquiry, the world is independent of our theories about it, so that even in the limit of inquiry there is a risk (though it may be exceedingly small) that our theories are incorrect. If this risk holds good, then sense can be given to the notion of a world whose structure does not match what our theories in the limit of inquiry assert about that structure.

The names for these two opposing positions are perhaps misleading and at the very least unfortunate. Realism was originally associated with something it no longer is associated with: Plato's theory of transcendent forms. The ancient Greek philosopher Plato held that the world of our ongoing sensory experience, the world of ordinary objects that we encounter through the medium of our biological senses, is not ultimately real. What is "really real" are the otherworldly **Forms** of which the ordinary objects of sensory experience are but imperfect copies. Hence, for a Platonist, reality is something beyond sensory experience, something utterly independent of it, something that sensory experience can only aspire to represent inaccurately, not actually constitute. Antirealism thus started out as the opposing view that the "really real" was to be found completely within the world of our sensory experience. It was the denial that there were facts of the matter outside and beyond all possible sensory experience. Over the ensuing 2,400 years since Plato this basic dispute has undergone a host of changes and subtle twists until it is now almost unrecognizable with respect to its original version. Nowadays it is antirealists who are taken to be advocating the controversial position, not realists. For example, realism is no longer associated with belief in a transcendent reality of some kind; instead, the realist insists on the theory-independence of reality. To be independent of theory is not the same as to transcend all possible experience. The former holds out the mere possibility that reality outruns our sensory experience, whereas the latter requires that it does so. Nowadays realists argue not that we could never know the truth but only that we are not guaranteed to find it eventually. We'll just have to wait and see how inquiry turns out. The antirealist, on the other hand, wants to guarantee that we eventually find the truth because the antirealist wishes to declare bogus the very concept of a world whose structural nature is independent of our theories in the limit of inquiry. For the antirealist, what the world is really like *must* be representable by our theories in the limit of inquiry; for the realist, there is no justification for putting the word 'must' into that sentence. Most likely we will figure out the nature of things, but there is no guarantee of this. Why not? Because the structural nature of things is independent of our theories—even our theories in the limit of inquiry—and therefore the risk of error in the limit, no matter how tiny it will no doubt become, will never drop to zero. Or, so says the realist.

Many of the above issues get entangled in a thicket of cross-reactions and ontological disputes whose relations to the original fight are not exactly clear. Often, people call themselves realists who defend the notion that the theoretical terms of a mature scientific theory refer to entities whose nature is independent of those theories; their opponents, sometimes called fictionalist antirealists, suggest the opposite, that theoretical terms are convenient fictions, of instrumental use only, useful for certain purposes but not to be taken literally. Hence, the idea is born that there is a realism/antirealism issue with respect to any form of discourse that appears on the surface to posit the independent existence of unobservable structures. One finds realists and antirealists fighting it out in mathematics, for instance. Numbers exist as independent entities, says the mathematical realist. Numbers are merely convenient concepts to be analyzed as something other than independently existing entities, says the mathematical antirealist. In psychology there is a dispute about whether the states of **folk psychology** exist as real structural entities independently of the folk-psychological discourse in which apparent reference to them occurs. Are there beliefs, desires, hopes, fears, and so on, or are those just useful concepts to be analyzed in some way as nonentities abstracted from a complicated form of discourse? Sometimes the realist side of a dispute likes to put its position by saying "there are facts of the matter." The antirealist side of the dispute can then be put by attributing to it the claim that "there are no facts of the matter, there are only" (to coin a phrase) "facts of the theory about the matter." What turn out to be "the facts" are partially constituted by and dependent on the humanly invented conceptual schemes embedded in the theories used to study the matter under question. Or, so says the antirealist.

Insofar as Peirce claimed that the structure of nature is *constituted by* what our scientific theories in the limit of inquiry assert about that structure he was an antirealist. He is sweeping away the very notion (on pragmatist grounds) of a structural reality whose nature is independent of our contingently evolved conceptual frameworks. Insofar as somebody such as myself denies that the facts of nature are constituted by what our theories say about things, I am a realist. Notice that I am not asserting something otherworldly like Plato did. That element of traditional realism turns out to be optional. I can be a realist and insist that reality is not transcendent in any mystical sense.

In the last two decades the realism/antirealism dispute has experienced a strong resurgence within philosophy. Some of the most intense discussion and debate has centered around a version of the dispute in which it is taken in a very general sense. This is unfortunate for us because we are interested in the dispute as it applies specifically to the philosophy of science—it is the scientific case in which we are interested. The result is that much of the discussion tends to be a mixture of very general considerations, considerations at the most abstract level of metaphysics, together with highly specific case illustrations from science. Many of the "big names" in contemporary philosophy of science have lined up to take sides in the dispute. Sometimes one gets the impression reading the massive literature on this dispute produced by these big names that the two sides are talking past each other, that they are not interested in the same issues in the end. I have heard realists declare the antirealist Micheal Dummett's arguments to be "utterly irrelevant" to what the real issues are. I have heard Dummettians declare that such realists don't understand what is at stake, what the real issues are. Bas van Fraassen has insisted that the main argument realists consider their most powerful weapon is simply a nonstarter, nothing more than a bizarre demand dressed up as something more substantial than it really is. I have heard realists insist that the antirealist Richard Rorty engages in underhanded realism, that he "sneaks" an independent external world back into the picture despite his clamoring to the contrary (so the charge here would be the serious one of inconsistency). I have heard an antirealist argue against realism on the ground that it is committed to what is ultimately a "spooky" view of nature, and that nature could not be spooky in the way required. Antirealists like to charge realists with being unable to escape something they call ultimate skepticism. The idea seems to be that, if there is a theory-independent world, we could never be in a position to know for sure, with absolute certainty, that what our theories say the structure of the world is correctly matches the structure of the world. Realists don't see what the problem is. Why in heaven's name would anybody ever think that science could deliver certain truth? It isn't in that business in the first place. It delivers highly probable truth. Hence, realists find the ultimate skepticism complaint to be a red herring issue. In the limit of inquiry we might have extremely good grounds for having extremely high confidence that our theories have got it basically correct. This tends to enrage antirealists, who believe that the realists are being strangely obtuse about the whole matter. Under these conditions the debate can degenerate into personal epithets. What tends to be revealed more than anything else in the realism/antirealism debate is the role of underlying personal temperament in the generation of deeply metaphysical disputation. It may be the case that reason and argument run out of gas on us here. Reason and argument cannot decide the issue because the issue is being fed by nonrational sources. But before we give up and declare it all to be a matter of temperament, let us canvass the main arguments that have been offered.

10.2 The Cosmic Coincidence Argument for Scientific Realism

It is a curiosity of the realism/antirealism dispute that the impetus for the *arguments* comes from one side: the realist side. The writings of antirealists generally consist of attempts to blunt arguments for realism, rather than of attempts to produce positive arguments in support of antirealism. Consequently, realism generally holds the defensive ground and antirealism takes the offensive tack. All the arguments for realism form one family of related arguments—they all pretty much work off the same basis. They are all arguments to the conclusion that realism is the only (or, at least, the best) explanation of a certain feature of scientific experience, a feature that is obvious to those who actually practice science on the inside, so to speak. The last point is critical. Often these arguments for realism leave nonpractitioners a bit cold—they don't see the force of them. Realists are inclined to attribute this to the outsider's lack of familiarity with the nitty-gritty details of everyday science. The nonscientific public has very little knowledge on the outside of what it is like to practice science from the inside. To the skeptical outsiders this sounds like special pleading on the part of the realists—and so the dispute spins on and on. I am going to explore two versions of the basic argument for scientific realism, the **convergence** argument and the **manipulability argument**. Both of these arguments are versions of what has come to be labeled the **cosmic coincidence argument** for scientific realism. As usual in philosophy, the name is unfortunate and a bit misleading. It sounds like the proponent of the argument is going to be appealing to something called cosmic coincidences when in fact the argument depends on rejecting the view that coincidences of "cosmic" proportions could ever occur; or, perhaps a weaker version is not that they could never occur, but that it would never be rational to believe that they occur. So, let us be clear here. The realist is going to be presuming that any view of science requiring coincidences of a cosmic proportion to occur must be an inadequate and false account of science. The charge will be that the antirealist view of science is committed to just such cosmic coincidences, and therefore it is to be rejected in favor of realism, which does not require such cosmic coincidences.

10.2.1 The Convergence Argument

The first version of the cosmic coincidence argument I will call the convergence argument. It is a quintessentially insider's argument, extremely impressive to practitioners, as against nonpractitioners, precisely because the former as opposed to the latter witness the alleged convergence first hand on a daily basis. Let R be a theory that attributes a host of unobservable properties to a host of theoretical entities. To have a specific case in front of us, let R be the current immunological theory of AIDS. Now, using R a practitioner can generate a large number of predictions about the outcomes of various observable experimental states of affairs. For example, if AIDS is caused by human immunodeficiency virus (HIV) particles fixing to a section of the CD4 surface protein on T helper cells and thereby invading those cells (and let us take all those events to be unobservable), then any drug that blocks the CD4 surface protein should lower the infected helper-cell count in an HIV-infected individual when that count is performed on a standard T cell smear (which is a perfectly observable procedure). Further, we could predict that, if R is the case, then any molecule that is structurally isomorphic to the CD4 surface protein (again, an entirely microstructural and unobservable state of affairs) should bind HIV particles in a standard immunoassay (a perfectly observable laboratory procedure). Further, any drug that washes away the CD4 surface protein from the surfaces of T helper cells (again, a theoretical reality) should render that class of immune cells immune to HIV infection as observed in a standard immune challenge experiment (another observable laboratory procedure). The list could be continued, but these three experiments should suffice to give the reader the general idea, which can be summarized in the following way. Practitioners design a host of *different* experiments that presuppose a certain theory. These different experiments are so designed that they can be repeated by any number of properly trained and equipped practitioners working in any number of different locations. The repetitions often amend the original design in creative ways so as to put the theory under further test. Then practitioners note a fascinating result: Everybody gets outcomes that are compatible rather than incompatible with the theory used to design the experiments. Everybody gets outcomes that approach a common result. The suggested outcomes of the three experiments I described above are all outcomes one would expect if indeed the current immunological theory of AIDS was basically correct. Any unexpected aspects to the outcomes involve errors of minor detail, not anything major. Call this feature whereby the outcomes of independent experiments all point to the same theoretical state of affairs the **convergence** (that is, the coming together) of experimental results. Now surely, argues the realist, it could not be a coincidence that the immune systems of AIDS patients behave experimentally, time and time again, under so many different conditions, *as if* the immunological theory of AIDS was correct, even though it is completely off the mark. To suppose that the underlying test theory is inaccurate would be to accept the convergence of all these experimental results as a coincidence of nearly cosmic proportion; in short, it would be a preposterous miracle that the experimental results converged on the same one apparent theoretical account. But, any view that implies the miraculous is surely unacceptable as philosophy (though perhaps it is not unacceptable as theology). Yet, the antirealist, who claims that we haven't justification for thinking that our theoretical posits are anything more than conceptual-scheme-dependent and convenient fictions, would seem to be committed to treating the convergence of observable experimental outcomes as utterly miraculous, as constituting a cosmic coincidence. After all, for the antirealist there is no reason to suppose that there "really" are T helper cells, that they "really" have CD4 protein molecules on their surfaces, and so on. These are merely instrumentally useful concepts dependent on the conceptual schemes currently used in immunology. If that is so, there would seem to be no grounds for why the results of experimentation converge in the way they do converge—the convergence is a preposterous coincidence, a staggering miracle, of cosmic magnitude. If theoretical terms were simply "made up" and did not actually refer to anything in the world, then observable experimental outcomes would more likely diverge, not converge. They would more likely spread out randomly across the spectrum of possible theories instead of clustering on the same or very similar theories.

Often realists like to generalize this convergence argument and talk about the success (meaning the experimental success) of science in general. That is, the reader should not assume that the convergence involves just one domain like immunology. This sort of convergence happens in every domain of scientific inquiry, all the time. Hence, we can construct what might be called the *ultimate* convergence argument (this name is due to the antirealist Bas van Fraassen). This ultimate argument can be summarized in a single sentence: *Realism is the only metaphysical position that does not make the general overall success of science a preposter-*

ous miracle/coincidence. It is implied in the sentence that nothing preposterous is to be accepted by a philosopher if there is a nonpreposterous alternative. Realism is that nonpreposterous alternative, for realism provides a nonmiraculous and noncoincidental explanation for the convergence of experimental outcomes: Specifically, our mature theories used to underwrite the experiments in question are basically correct and there really are the entities and properties in the world that those theories posit. Thus, realism is justified as the preferred philosophy of science over antirealism. The realist philosopher of science Richard Boyd summarizes the ultimate version of the convergence argument succinctly and precisely when he writes that realism "provides the only scientifically plausible explanation for the instrumental reliability of the scientific method." Boyd's description has the virtue of filling in a bit the sense of success at work in the argument: It is the fact that the methods used in attempts to intervene in and manipulate the outcome of the course of natural events are *reliable* methods that counts as our being successful in applying our theories to nature. It is not merely that the convergence happens occasionally, but that it happens time and time again to the point that the outcomes can be depended on and trusted as being stable. Here one can see the connection to Boyd's notion of homeostatic property clusters, which we investigated in chapter 6. Success can be taken to mean that the taxonomies of our mature scientific theories must be, for the most part, correctly cutting nature at its seams so as to be revealing for our potential manipulation the objectively existing homeostatic property clusters in the universe.

It is important to realize that the convergence argument has a quantitative component to it. If the amount of convergence was modest, then the argument would have little force. What gives the argument its tremendous power is the massiveness of the amount of convergence in science, an amount that many "outsiders" simply have no clue obtains. That massiveness to the convergent success tends to be underappreciated. The reader might recall that the feminist science critic Lynn Hankinson Nelson argued that those of her sister feminist science critics who are fond of relativism (which is virtually all of them) fail to realize how implausible relativism is in the face of the massive convergent success of science. It would be one thing if there were basically only one experimental way to test some theoretical account of the structure of a certain phenomenon. The fact that repeated runs of this same one experiment kept producing the same results would not be especially impressive in that case. But in fact there are usually myriad ways to test a theoretical account of the structure of some phenomenon. If the outcomes of 27 different (that is, not simply repeats of the same one experiment) experiments all converge on some theory R, or 37, or 237, then it becomes harder and harder to adopt an antirealist posture about theory R, it becomes harder and harder to be an instrumentalist about theory R. At a certain point, the massiveness of the convergence becomes sufficient to destroy the antirealist position as a live option. There are no general rules about when that point is reached. Figure 10.1 illustrates how the convergence argument might

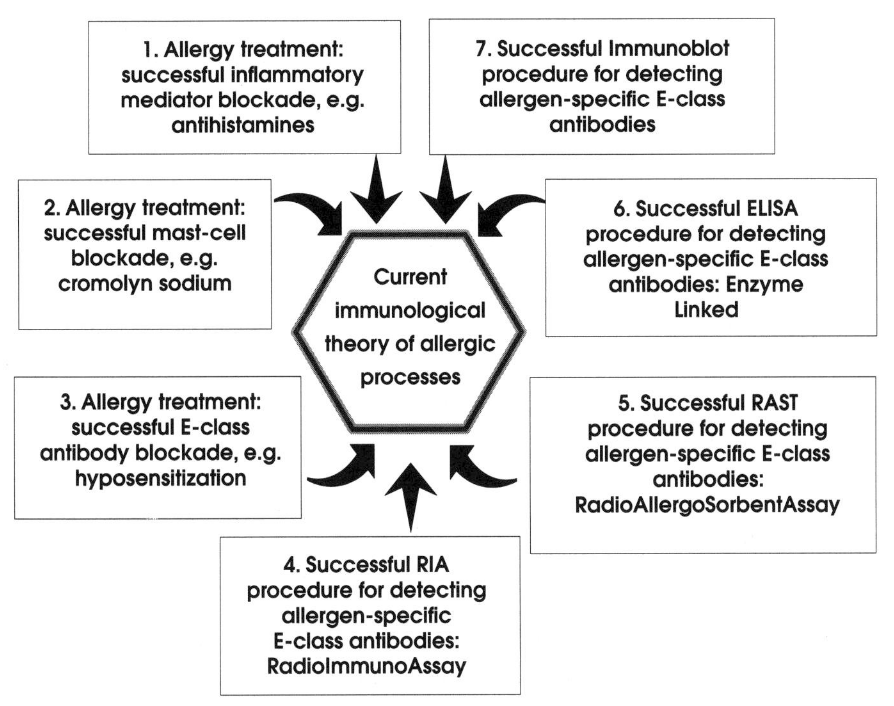

Figure 10.1. An example of convergence in immunology.

be applied to the case of the immunological theory of allergies, a theory whose main outlines we surveyed in chapter 1. The figure shows that over the course of time an impressive number of theoretically independent experimental procedures have been developed—procedures that depend on different background theories in different fields collateral to immunology—all of which produce observational outcomes that converge on the theory that allergies are due to the events related in detail in chapter 1. If the current immunological account of allergic disease depicted there was mostly false, then so many different manipulative experimental techniques converging on the same theoretical outcome would be a preposterous coincidence of cosmic proportions. The only plausible explanation of the convergence is that the current immunological account of allergies is mostly correct.

How antirealists respond to the convergence argument depends on whether they are social constructivists or not. Social constructivists like Bloor, Latour, Woolgar, Shapin, Schaffer, Harding, and Longino, as we saw in the previous two chapters, argue that the convergence is not genuine convergence. It is an artificial result due to the tendency of theories, social metaphors, and projected psychological biases to be self-affirming when they are used to evaluate their own products. The idea is to grant the reality of the convergence as a superficial phenomenon but to argue that its cause is not what the realist wishes to claim it is: Its cause is not that the content of science correctly represents some inquiry-independent reality. I argued in the previous two chapters that this social constructivist form of antirealism is inherently implausible and based on poor argumentation, and I will not repeat my critique here. Instead I want to look at how an antirealist who is not a social constructivist might respond to the convergence argument for scientific antirealism.

The antirealist Bas van Fraassen has responded to the convergence argument by insisting that the success of science is *not* something that requires any kind of explanation. It is no business of the philosophy of science, argues van Fraassen, to give a metaphysical account of why science enjoys the success it does. We do not demand this of any other successful institution; and, besides, it is impossible to fulfill. Therefore it is unfair of realists to ask for an explanation of science's success so far. Van Fraassen is worried that the convergence argument is some kind of throwback to "medieval" ways of thinking in which one foolishly requires that "the order of human thought match the order of nature itself." It is true, van Fraassen argues, that many scientific theories work quite well, but that is not due to some generalizable feature they all share, nor is it due to some happy correspondence between the structure of external nature and the structure of humanly contrived categories of thought; it is just a brute success, and van Fraassen argues that we should be mature enough to accept it on its own terms without requiring that some further account be given of it. This is an interesting argumentative move by van Fraassen, for it should remind the reader a little of one of my own concerns in chapters 8 and 9. What van Fraassen is worried about is that requiring a match between our theories and nature flirts with "science by fantasy," that it encourages us to believe that we can think nature to be what it is, that we get to make it all up in the course of making up our theories. In chapters 8 and 9 I was concerned that the social constructivists we surveyed, particularly the two feminists Sandra Harding and Helen Longino, were in danger of promoting such a view of science. This is a curious situation. I accused certain social constructivist antirealists of advocating the practicing of science by conventional decision, whereas van Fraassen is worried that realism is going to end up committed to that practice. One of us must have a misguided worry. I suggest that it is van Fraassen who does. The convergence argument does not, as van Fraassen assumes, make it necessary that there is a happy match between the order of human thought and the order of nature—it presupposes only that such a match is the best explanation of something for which we otherwise have absolutely no explanation. The matching, if it occurs at all, can be thought of as entirely accidental, as a stroke of dumb luck on our part, if that would make van Fraassen feel better about the whole matter. On the contrary, as we have seen with Peirce, it is usually the antirealist who demands that the order of human thought match the order of nature, for it is the antirealist who *defines* reality as what our theories license as appearance in the limit of inquiry. For the antirealist reality collapses into appearance in the limit of inquiry. It is the realist who holds open the *possibility* that even in the limit of inquiry our theories might have the story of reality incorrect in certain ways.

Van Fraassen claims that it is no business of ours to give an explanation of the general success of science. But why not? Why must this admittedly impressive level of success in science remain a brute inexplicable aspect of it? Well, argues van Fraassen, mice are pretty successful at avoiding being caught by cats, but it would be absurd to explain this success by saying that mice have a correct theoretical understanding of cats—their dangerous behavior, where they hang out, when they move about, how they like to stalk prey. Mice are successful at avoiding being caught by cats without having any beliefs about them at all, seems to be the suggestion, for mice who were inept at avoiding cats were selected against in the course of evolution—only good cat-avoiding mice survived. Accordingly, we do not explain the success of mice at avoiding being caught by cats by attributing to mice correct mental representations of "cat reality" (except, of course, in anthropomorphic cartoons about cats who fail to catch savvy mice). By analogy, van Fraassen argues, mature scientific theories are successful because unsuccessful theories lost out in the competition against them, not because the mature theories correctly match the structure of the world in some way, not because they represent reality accurately.

One response on behalf of realism is provided by Philip Kitcher. Kitcher argues that van Fraassen's analogy to evolutionary accounts of mice is too shallow. Van Fraassen makes it sound like the evolutionary biologist rests content with noting how mice adept at avoiding cats were selected for over mice not so adept at avoiding cats. This is simply not true, Kitcher argues. The evolutionary biologist is interested in discovering what general features of the selected for mice are causally responsible for their greater adeptness at avoiding cats. More details are forthcoming than merely noting their greater adeptness. Is it because the selected for mice have keener hearing than the mice not selected for? Is it the possession of foot pads more sensitive to vibration that accounts for the greater adeptness at avoiding cats? By analogy, the philosopher of science does not rest content with what van Fraassen rests content with: simply noting the survival of empirically adequate theories. We want to know what general features of scientific theories account for their survival against their competitor theories. Is it something about their explanatory structure? Is it that the successful theories have taxonomies that map onto real distinctions among the entities investigated? Is it that manipulation of experimental systems while using those theories is more sophisticated? Answers to such questions as these fill in the story of why successful theories survived, and the answers fill in that story without postulating a magical match between the order of nature and the order of thought.

A second response to van Fraassen's argument is to question van Fraassen's reading of his own analogy. A case could be made that, in a certain sense, mice do accurately represent cat-reality. The representation is not semantic but anatomic. The selection pressures of its evolutionary history have honed the mouse body into a device adept at avoiding being caught by cats. Given the properties of cats (fast, keen eyesight in the dark, a svelte and noiseless step), mice physiology was thereby causally constrained and shunted down a particular evolutionary path. The nervousness of mice in a sense "represents" the fastness of cats (for it reproductively benefitted mice to be extra vigilant for approaching predators such as cats). The physical aspects of the skeleton and muscles of mice, which give them their small size and quickness, in a sense, "represent" the keen eyesight and svelte step of cats (for it reproductively benefitted mice to be able to remain undetected by prowling cats and to retreat quickly when detected by cats). In the case of mice, the representation of cat-reality is not abstracted from the physical being of the mice and set up as an independent semantic entity; in the case of a scientific theory, the representation of reality is put into an abstract and semantic form. Looked at in this way, van Fraassen's analogy backfires on him—it proves the very opposite of what he thought it showed. We can explain the success mice have at avoiding being caught by cats by appealing to how evolution has honed the physiology of mice into a kind of accurate representation of cat-reality.

I am sure that van Fraassen would be unimpressed by my argument. He would consider it too far-fetched to be taken seriously. His writings make it clear that he would consider the attempt to read cat-reality into the anatomical structure of mice nothing more than a sign of how embarrassingly desperate realists can become in the service of pushing their pet view of science. Perhaps we have hit the ground of sheer temperamental differences here. Antirealists like van Fraassen find it bizarre that realists seek to explain something that, to a person of his temperament, just ought to be accepted as a brute fact: Our mature theories allow us to intervene successfully in nature. Realists find it bizarre that a fellow like van Fraassen fails to hear the success of science crying out for explanation—how could he be so deaf, to speak metaphorically. I say to-may-toe, you say to-mah-toe. What sort of *argument* could sway anybody in this situation? Probably argument is going to amount at most to preaching to the choir. The realist is addressing other realists in her arguments and the antirealist is addressing other antirealists in her arguments. There is an aspect of preaching to the already converted that infests the whole debate, an aspect especially highlighted by the second argument for realism that we shall examine.

10.2.2 The Manipulability Argument

The philosopher of science Ian Hacking has argued that we ought to pay some attention to how scientific practitioners themselves understand the ontological consequences of what they do in their laboratories. Hacking has been complaining for some time that traditional philosophy of science has overstressed theories at the expense of ignoring the role of experimentation in science. Philosophers have a tendency to concentrate on the finished product of mature science: the pristinely organized theory as presented in a textbook. Real day-to-day science often has little to do with such textbook theories, as Kuhn pointed out long ago. Scientific practitioners live and die by the worth and importance of their experimental achievements. What really matters to practitioners, argues

Hacking, is their success or failure in their attempts to manipulate experimentally the important entities and properties "out there" in the domain of nature that they investigate. A goodly proportion of those important entities and properties are unobservable ones. Does their unobservability, their theoreticalness, strike practitioners as a highly problematic, an epistemologically controversial, aspect of them? Not very often, Hacking reports. Only crazy philosophers who have read too much Descartes think that the unobservability of these entities and properties poses some special problem. The average experimental scientist has no more doubts about the reality of theoretical entities and properties that she or he has successfully manipulated, time and time again, under many different conditions, than she or he does about the ordinary everyday macroscopic objects of obvious reality in the local environment. Why do they have no more doubt about the one than the other? Because, says Hacking *successful interference with nature is the stuff of reality*. We must be careful here, argues Hacking. Practitioners do make a very critical distinction between the entities and properties in question and their *current theories about* such entities and properties. Many practitioners are highly skeptical that the theoretical accounts of the nature of the entities and properties they successfully manipulate in their experiments are totally accurate accounts, but they have no doubts that there are entities and properties denoted by the theoretical terms in question. For example, no immunologist seriously doubts the reality of E-class antibodies; but about the current theory which purports to give an account of the nature of E-class antibodies—their structure, their causal behavior, and so forth—immunologists may have all kinds of doubts. In short, actual practitioners, argues Hacking, tend to be realists about *entities* but skeptics (and possibly antirealists) about *theories*. The traditional realism/antirealism debate runs these two issues together when they ought to be distinguished. How did Hacking arrive at this view? He reports an interesting personal experience in a physics laboratory. He had been granted permission to follow around a bunch of physicists who were in the process of carrying out an experiment. Hacking watched the physicists and asked questions like "what are you doing now?," and "what does that green light on the console mean?" The physicists patiently answered his questions. Near the end of the setting up procedure for the experiment, Hacking watched the physicists manipulate a real doohickey of a device (Hacking describes the device, but to a nonphysicist it sounds like something out of a sci-fi film). The device contained a moving drum, a cylinder that rotated on some bearings. One of the physicists pointed another smaller device at the surface of the drum and Hacking began to hear a funny hissing noise. The physicist slowly rotated the drum underneath the smaller device from which the hissing sound was coming. In answer to Hacking's question about what was going on, the physicist floored Hacking with the nonchalant reply "I'm spraying the drum surface with electrons." Hacking was floored because the tone of voice of the physicist made it abundantly clear to Hacking that the physicist no more doubted the reality of electrons than someone would doubt the reality of the earth, of the food they eat, of the clothes they wear; yet, electrons were, in the days of the positivists, often used as a standard example of unobservable entities with a questionable ontological status. Philosophers have traditionally taken subatomic particles like electrons to be paradigm cases of theoretical entities about which someone could reasonably be an antirealist. After regaining his philosophical composure, Hacking informed the physicist of how incredible his nonchalant reply had been, of how many philosophers would hoot and holler at what the physicist had said. "How do you know that you were spraying electrons on the surface of the drum?" Hacking told the physicist those hooting philosophers would ask of him, "You didn't see them or smell them, you just pressed a trigger on the second smaller doohickey and a hissing noise was heard—you inferred everything else including the conclusion that you were spraying the drum surface with electrons." "Listen," the irritated physicist replied to Hacking, "*if you can spray them on, then they're real!*"

Hacking reports having undergone a gestalt switch at that moment, a conversion experience, an "aha!" experience. Of course! How simple and commonsensical could it be? *If you can spray them on, then they're real.* Hacking realized that this was nothing unique about spraying as opposed to any other form of manipulation. Thus, a general argument was abstractable from the physicist's brutally honest and unrehearsed reply:

for any putative entities x, if one can successfully manipulate x's, then x's exist.

The physicist's response might sound to certain oversophisticated ears like a superficial knee-jerk remark, but it harbors a deep, deep, truth, argues Hacking, the same truth at the heart of the average person's realism about the everyday macroscopic entities in the local environment. It would surely be peculiar, but suppose you were asked to provide an argument, a justification, for the belief that the book you are now reading is "real." It would be a mistake to seize on some *one* aspect of it—say, the fact that you seem able to see it clearly or feel it in your hands—for one particular sensory experience can be illusory or distorted in various ways. A much more powerful argument would be one that appealed to your ability to manipulate the book successfully in a variety of ways under a variety of altered conditions. You want to sit on the book? Do it. There, see? Success. You want to throw the book at the wall? Go ahead. More success. Now, use the book to scratch your ankle. Good. Now, if the book was not real, if it did not have a relatively stable material structure that you have some fairly accurate understanding of, your ability to meet with all this success would be, shall we say, a bit mysterious? But the reality of the book removes the reputed mystery. The book is not a mere convenient fiction, it is not a mere invention of the conceptual scheme you operate with; rather, it has a structural nature independent of that conceptual scheme, a nature whose independence is called on to explain your success at manipulating the book experimentally.

One obvious problem with Hacking's manipulability argument is that in many mature sciences experimental manipulation depends on the use of complex devices whose construction already presupposes a host of other theories, and therefore such manipulation already presupposes the existence of many other unobservable entities posited in those other theories; in other words, the experimental manipulation only counts as successful on the condition that a host of further theoretical conjectures are assumed to be true, that a host of other unobservable entities exist. Technological devices and procedures do not grow on trees, as it were. Practitioners build devices using their theoretical beliefs to design them. Accordingly, if those theoretical beliefs are false (say, because there do not in fact exist the unobservable entities the beliefs are about), then the devices do not necessarily function like their designers think they do. Suppose I want to manipulate some putative unobservable entities called blickledorps. I construct an experiment to do so, but the experiment requires the use of a fancy machine called a fungometer. According to certain other theories I hold, fungometers depend for their proper functioning on the existence of other unobservable entities called fungos, as well as on the existence of yet another class of unobservable entities called ferples. Let us suppose that the fungometer itself contains a subpart called a schlunkmucker whose function it is to help manipulate ferples by manipulating yet another bunch of unobservable entities called schlunks. And this may continue on and on. Nature is a tightly woven fabric, and its structural parts interact in a close and incestuous way. Rarely will a practitioner of a science be able to manipulate one kind of unobservable entity without having to mess around with at least one other kind in the process. Hacking is aware of this criticism, but his response to it seems a bit weak. He argues that the manipulation of the unobservable entities of interest will, in many cases, not involve presupposing the existence of yet other unobservable entities, but rather the use of simple "home truths" and further theoretical beliefs whose epistemological warrant is relatively secure on other grounds. But this seems to be just plain false. Recall the esoteric details of the ^{51}chromium release experiment we surveyed in chapter 4. A scintillation counter was required to measure the amount of radioactive ^{51}chromium in the supernatant. That experimental situation is hopelessly theoretical: To detect observably the alleged cancer-killing properties of the alleged NK cells that we are supposedly manipulating in the experiment we must use a device whose very construction presupposes nearly the whole of quantum physics. Far from being unusual, the worry is that this situation represents the typical case. Where Hacking gets his idea that only simple "home truths" and independently established theoretical claims hover in the background I do not know. It seems a bit made up and rather a case of wishful thinking—especially if we are drawn to any sort of Quinean holistic model of theories. If theories have an irreducibly holistic aspect to them, then we ought to suspect that it is not going to be easy to design experiments in which manipulation involves only the unobservable entities of interest to us and no other ones.

Certainly the presupposition of the existence of other unobservable entities weakens Hacking's general argument for realism. The worry is that the realist practitioner only *thinks* she has succeeded in manipulating the unobservables of interest to her when in fact so much speculative theory underwrites so much of the experimental apparatus, that she is simply mistaken; a kind of delusion has been produced due to the massive edifice of theory involved in the experiment. To drive the point home, we might imagine a naive pseudoscientist who argues for the existence of his favorite pseudoscientific entities using a Hacking-style argument: He cites his "success" at manipulating the alleged entities when in fact his experiments are so hopelessly bound up by presupposition with other equally theoretical beliefs in other equally questionable unobservable entities as to produce the delusion of successful manipulation. Until this worry is removed, Hacking's manipulability argument will suffer from this significant internal weakness.

Another criticism of Hacking's argument has been put forward by Hilary Putnam. Putnam objects that Hacking argues X's exist on the basis of having successfully manipulated X's, but that Hacking does not say what the existence claim *means* in such a case. Hacking conveniently borrows the assertion of existence as though that assertion doesn't carry with it a host of conceptual implications. Existing *things*, argues Putnam, have spatiotemporal locations, are countable, and so on; this creates a problem for Hacking because his example of spraying the electrons on the drum involves subatomic particles whose thinglike status is very much in question under quantum theory. Subatomic particles have no precise spatiotemporal locations, for they are probabilistic entities, half-wave, half-particle. So, what does it mean to say that they exist? It looks as though Putnam fails to appreciate Hacking's attempt to distinguish between theories of X's and X's. It is likely that Hacking could grant all of Putnam's worries but argue that the nature of electrons—what they are like as things or nonthings—is precisely what you need a *theory* about electrons for. Such theories are always tentative, subject to change, and loaded with errors of detail. Hacking's position is that electrons, whatever their nature is under some to-be-discovered theory, exist because, whatever their nature is under some to-be-discovered theory, we can somehow successfully manipulate them. Putnam objects to the very language used in such a reply, for electrons may not be thinglike enough to justify use of the standard pronouns 'they' and 'them' to refer to electrons in the course of asserting that "they exist because we can successfully manipulate them." Putnam's complaints seem a bit strained. In other contexts besides quantum physics we allow people to make reference to entities whose structural nature is problematic using the standard pronouns. For example, in folk psychology theorists often find it useful to speak of beliefs, desires, hopes, fears, and so on, as if they were thinglike entities inside of people when there are lots of arguments around that they are not any kind of things at all. It can be a *discovery* that what we have been talking about all along with phrases like 'those X's' are not thinglike entities at all, yet Putnam's position would seem to rule this out as a coherent possibility. Hacking's manipulability argument, as we have seen, has problems, but the one Putnam argues for is not among them.

10.3 The Case for Scientific Antirealism

Since the demise of the logical positivists one does not find many antirealists among the total population of philosophers of science. Quine taught us how to reject the epistemology of positivism and still be realists of a sufficiently robust sort. The prevailing view seems to be that antirealism made a lot more sense, had more plausibility, when science was younger and less mature than it is now, when it was more ill-formed and contained many more gaps than it does now. As science has progressed, as its general success has increased over the years, serious skepticism and antirealism about its substantive results has slowly come to seem more and more philosophically sophomoric, as though it would be an act of mere skeptical petulance to persist in doubting the overall theoretical accuracy of science. It came to seem as though there was something conceptually strained about antirealism, that it was some sort of academic stretch, the taking of a contrary position for the mere sake of being contrary. The suspicion was born that a serious antirealist is a little bit like the proverbial "village atheist," somebody addicted to the high obtained from the perverse pride of being the only out-of-step "free-thinking" person in their village. The antirealist wants to be different, not because the current burden of proof suggests that being different is being correct, but because to be different is, well, simply to be different.

In recent years the status of antirealism has begun to change toward more respectability. This has been due to the work of four centers of philosophical reflection on science: the "school" of social constructivist critics of science, the work of the empiricist philosopher of science Bas van Fraassen, the work of the philosopher of science Larry Laudan, and the work of the philosopher of science Arthur Fine. We shall begin our investigation of antirealism with a deconstructionist argument for antirealism.

10.3.1 Antirealism: Success Is Self-defeating for Realism

During the past quarter-century an intellectual movement of sorts arose in the literary disciplines that came to be known as **deconstructionism**. While it has been most influential in literary criticism, certain French intellectuals who are classified as philosophers took up the cause of this movement. The result has been a bit odd. While deconstructionism spread like a roaring fire through the field of literary criticism and assorted regions of the social sciences, it has been met largely with indifference or contempt by Anglo-American philosophers. The devotees of deconstructionism, as might be anticipated, interpret that indifference and contempt very differently from the way Anglo-American philosophers who practice it do. To the former the indifference is based on misunderstanding, fear of the new, and commitment to intellectual methods and proclivities that are directly threatened by the deconstructionist way of doing things. To the latter the indifference and contempt is justified indifference and contempt, for deconstructionism to most Anglo-American philosophers is a faddish and confused mishmash of incoherent literary criticism. The Anglo-American philosopher of language John Searle published in the early 1980s a devastatingly negative critique of the work of the French deconstructionist Jacques Derrida that expressed the general belief among Anglo-American philosophers that deconstructionism can be properly disposed of as an incoherent and therefore pointless French game of witty word-twisting. The deconstructionists are, of course, unimpressed by all this, for they simply proceed to deconstruct all attempted refutations of deconstructionism, thereby showing how allegedly phony each of them is.

Steve Fuller is an Anglo-American philosopher who is nevertheless somewhat sympathetic to deconstructionism. He argues that each side in the realism/antirealism dispute actually, ironically, presupposes the truth of the other side. This is just the sort of thing deconstructionists are wont to claim about allegedly irreconcilable intellectual conflicts. If Fuller is correct, then the dispute is in a very strong sense an incoherent one, for one of the two sides wins only if the other side wins; but, because they can't both win (for they make directly contradictory assertions), neither of them can win. You see, it's all just a tangle, a thicket, of words. There is no reality beyond our descriptions, beyond our modes of speaking and writing. The substantive content of science is the same as the humanly composed text of science. Science is not about a world independent of the text of science. There are no text-independent facts of the matter.

Is Fuller correct that each side in the realism/antirealism dispute presupposes the truth of the other side? In particular, how is it that realism presupposes the truth of antirealism? Here is how, says Fuller: Realism is based on a kind of "credit." That is, the scientist is allowed to posit an unobservable entity or property subject to the understanding that eventually, even if it takes a couple hundred years of technological development, it will become an observable entity or property. When this "promise" is not made good on, argues Fuller (claiming merely to be following the reasoning of the realist philosopher of science Rom Harre), when the debt is not paid off in this way, then the purported entity or property justly fades away within science—it ceases to be posited and believed in. Fuller adds that this promise that all unobservables will eventually become observables could be a methodological requirement only on the assumption that antirealism is true, that is, the only justification one could give for such a requirement is that *the real is the observable in the long run*, and this is just what antirealism entails according to Fuller.

The problem with Fuller's attempted deconstruction of scientific realism is that it mistakes what is a matter of hope for a point of required method. A practitioner may no doubt hope or fantasize that a posited unobservable will someday become observable, but its becoming so is not a requirement, not something that in itself necessitates the rejection of the posited entity or property. Some posited unobservables are *intrinsically* unobservable, such as planets orbiting stars at spacelike separation from the earth, various mathematical properties and entities such as sets and numbers, certain large-scale phenomena like social classes, and certain microphenomena so exotic and remote from us as to be nearly insubstantial in character (for example, the charm property of quarks). Will anyone ever observe, in live-time as it occurs, a mutation in the genetic material of a living organism? Such an unobservable state of affairs no amount of technological development may ever render observable. It seems extremely implausible to suppose that should they remain forever unobservable, genetic mutations must therefore fade out of biological science eventually. Theoretical posits can earn their keep in our scientific theories in ways other than by being observable—they can do explanatory work, they can do predictive work, they can help to systematize other beliefs, and so on. It is a misrepresentation of realism to argue that it must be grafted to a severe form of empiricist epistemology to be legitimate, yet this is precisely what Fuller's argument does.

There is another increasingly popular argument against realism that does not show its deconstructionist origins so obviously. Some social constructivists argue that what we have called the cosmic coincidence argument presupposes a notion of success (convergence) that is bogus—theories are not successful in the way the convergence argument suggests that they are. The realists help themselves to a level and kind of success that is largely mythical, seems to be the argument. It is mythical because the success is question-begging: The same theory must be used to determine whether its own use meets with success. The social constructivist Harry Collins calls a version of this argument the **experimenter's regress**. Suppose, argues Collins, that a scientist makes a claim C using a certain theory T, which is then put to the test. The scientist builds a device to test C. Now, fancy devices need to be calibrated, they need to be tuned so that we shall know if and when they are working properly. Here is where the problem allegedly lies: For theory T must be used to calibrate the device, which will then be used to test claim C, which itself is taken as a test of theory T. We are caught, argues Collins, in a circle, a regress, which we cannot break out of. The theory used to calibrate the device is the the theory under test. So, it ought to be no surprise that the results we get from our tests appear to be successful results, for we are using a theory to test itself, and theories always give themselves high marks. The alleged massive success of science is a phony success, according to the argument, because it is a self-fulfilling and question-begging success.

Fortunately for the realist, Collins's argument is a bit on the short-sighted side. Rarely does it occur in science that the exact same theory T must be used to calibrate a device that will then be used to test theory T. Much more frequently it happens that logically independent theories are involved in calibrating test devices. As an illustration, recall the ^{51}chromium release assay we detailed in chapter 4. There, to test the claim that the immune systems of mammals contain NK cells capable of killing cancer cells without prior sensitization to them, we had to use a scintillation counter to measure the radioactivity in the supernatant after incubation of the target cells with the NK cells. Scintillation counters are constructed using the current theories of quantum physics and their calibration does not presuppose the truth of the immunological theory under test in the ^{51}chromium release assay. Further, the design, calibration, and operation of the centrifuge used to separate the mononuclear component of the immune cells from other parts of the blood serum presupposes theories that are logically independent of the one under test. Mature science is rather more "incestuous," with different fields of inquiry intersecting each other, than Collins's experimenter's regress argument allows. It is no wonder then that social constructivists prefer to focus on the very early days of a domain of inquiry, where occasionally one does find the situation Collins worries about (Shapin and Schaffer argue that Boyle had to presuppose his own theory about the physics of air to calibrate the air-pump he used to test that same theory). The problem with such a focus is that it elevates the nontypical case in science to the status of being the typical case—which it isn't. The scientific realist, accordingly, need not fear that the experimenter's regress is a serious and daily impasse in mature sciences.

10.3.2 Antirealism: Success Needs No Explanation at All

Bas van Fraassen, perhaps the most prominent antirealist philosopher of science today, is concerned to show that antirealism can be made into more than a misguided contrariness based on Cartesian epistemology—it can be made as philosophically defensible and respectable as realism. Van Fraassen is a philosopher of physics. This ought to send a warning flag up the flagpole, for if there is one place in physical science where antirealism has any chance of finding a home it would be in quantum physics. Indeed, van Fraassen pins the crux of his antirealist position on quantum physics; it is made to carry the argument. I mention this because I believe it to be a very questionable strategy. No other area within science is as controversial, as rife with unclarity, and as conceptually conflicted, as is quantum physics. A case could be made that *nobody* really knows what in tarnation is going on at the quantum level of physical structure. No one in their right mind doubts that the formalism of quantum physics works—its equations can be used to interfere successfully in nature—but providing an understandable conceptual interpretation of what the formulae mean is another matter. Given this conceptually confused state of affairs, it seems precarious to ask quantum physics to bear heavy philosophical weight in the course of an argument for a metaphysical position, but that is just what van Fraassen does.

Like any crafty reasoner van Fraassen makes sure to attack the opposition first before presenting his own positive replacement doctrine. First he shall show us how untenable realism is, and only then will we be invited to consider the cogency of his antirealist alternative view. Realism, says van Fraassen, is committed to the principle that the aim of science is to arrive at, and here we must be careful to quote him, a "literally true" story of how things are. Realists are right to be suspicious of the adverb in the phrase, for van Fraassen argues that almost no scientific accounts of any phenomena are *literally* true. He gives examples. 'Electrons orbit the nucleus' is a typical proposition one might find printed in a high school science text, the sort of sentence uttered by the narrator of a science documentary on television, but literally construed it is of course false. Electrons are not literally solid entities at a precise location and which move in an orbital motion around the nucleus. The sentence must be understood metaphorically to be taken as an accurate claim: Electrons are in some respects (but not in other respects) *like* "little planets," and their motions are in some respects (but not in other respects) *like* the elliptical motions of a real planet around a star. To invent a further example from our own domain of illustration, consider the proposition 'antibodies recognize their corresponding antigens and attack them'. This is just the sort of assertion that is by now commonsense doctrine in immunology. Interpreted *literally*, of course, it is false and misleading, for antibodies do not literally recognize anything. Recognition is a cognitive capacity of higher organisms capable of consciousness, and antibodies are nonconscious protein molecules. Further, to attack something in the literal sense would require the ability to form the *intention* to do that something harm. It is an intentional state, and again, only higher organisms capable of intentional mental states can experience them, not noncognitive, nonorganisms like antibody molecules. The proposition must be understood metaphorically if it is to be granted accuracy. Antibodies are so structured that they fit certain other molecules (antigens) in a way that immunologists anthropomorphically compare to what happens when one person recognizes another person. The behavior of antibodies is therefore similar in a few respects (but not in many others) to the behavior of persons. Antibodies are not literally like "little policepersons" who "attack" the "bad guys" (antigens), but it can be pragmatically useful to think of them in that way for certain experimental and explanatory purposes.

Van Fraassen claims that the use of metaphorical claims is a feature generalizable across all domains of science. Almost no scientific claims are literally true. How does this hurt realism? It hurts realism because van Fraassen explicitly rules out as a possible realist position the claim that science aims at providing, not a literally true account, but a metaphorically true account of phenomena. Metaphorical accounts of phenomena do not tell us what they "really" are like, hence it can't be a realist position to adopt such a goal for science. So, the realist can not defend realism by saying something like, "okay van Fraassen, you have a point—scientific assertions must be properly interpreted to be true, not taken literally." That response, according to van Fraassen, strips the realist position of its distinctive mark as a realist one. That scientific assertions are no more than metaphorical is just what the antirealists have been maintaining for the last four centuries. Or, so claims van Fraassen, for he characterizes the essential feature of scientific antirealism as a commitment to the view that the aim of science is to provide an account of things that is "empirically adequate," not literally true. Empirical adequacy is not the same as truth. False stories can nevertheless be empirically adequate stories. Empirical adequacy requires only that there be at least one formal semantic **model/interpretation** of the theory under which all its observational sentences are true, that the theory, to use the famous phrase from the works of Plato, "saves the appearances." In this antirealist picture of science, science is required only to keep up with the appearances, and the whole issue of a reality that might be different from those appearances can be left to wither away as the philosophically obnoxious notion it has always been, despite the efforts of generations of realists to civilize it and make it into a philosophically polite notion. It is a deep and important truth that a false theoretical account of a phenomenon can nevertheless be completely consistent with its observable appearances. The history of science is full of such cases, so full that this feature of our epistemological lot in life was openly recognized by philosophers of science and dubbed the underdetermination of theory by observational data. We spent chapter 4 exploring some of the ramifications of the underdetermination of theory by observational data. Van Fraassen seems to think that one ramification it ought to have is to render realism less plausible and antirealism more plausible. He wants to charge the realist with demanding of science something it is not capable of delivering: the truth. Insofar as realism demands that science cough up the truth about things, realism puts unreasonable pressure on science, realism foists overambitious expectations on science. Antirealism is a much humbler view, van Fraassen argues, and therefore it is a better view. Antirealism takes the heat off the scientific practitioner and lets her go about experimental business without the nerve-wracking weight of having to discover "reality" in the course of her work. The practitioner is free to play with concepts, to alter equipment, to try out "crazy" ideas, and so forth; for all the practitioner is required to do is to produce a theory all of whose observational sentences come out true on some interpretation of them. There is no requirement to "penetrate to the essence of things," "ferret out the real story of things," "unveil the metaphysical Being of things," or engage in any other kind of highbrow realist metaphysics.

Is this a fair portrait of what realism demands of science? Sure it is. Furthermore, there are solid grounds for thinking that science is perfectly capable of delivering on these demands. Van Fraassen's nay-saying to the contrary appears to be undermotivated. One suspects that van Fraassen has been misled by his overconcentration on quantum physics, where perhaps it makes sense to stick only with appearances and forego all hope of reaching truth. But that attitude seems not to fit very well into the rest of the vast terrain of science. It seems a bit perverse to tell the immunologist that he is not to reach for truth in his theories, that it would be an act of self-defeating hubris to try to do so. Far from it, the immunologist can point to a great deal of past success at reaching the truth about immunological matters of fact. The ontological and epistemological humility quantum physics breeds in the philosopher of physics is utterly uncalled for in certain other domains.

Philip Kitcher argues that the epistemological humility that motivates van Fraassen's antirealism—a humility that implies that realism is too ambitious as a philosophy of science—boils down to the worry that any preferred realist explanation of our success might be incorrect. Kitcher notes that the concept of incorrectness in this context is a relative notion: A given explanation of our success can only be taken as incorrect with respect to some competing explanation that is supposedly better. Therefore, Kitcher argues, the issue is not the mere possibility of incorrectness in the explanation of our success, rather the issue is the evaluation of rival explanations of our success. Kitcher argues that van Fraassen fails to see that the only standards obtainable for evaluating rival explanations of our success come from our general scientific picture of the world, and that in that picture there is no basis for making the sort of distinction between observables and unobservables that van Fraassen needs and wants to make. In our general scientific picture of the world we freely intertwine observables and unobservables. Richard Boyd makes a closely related point when he argues that consistency with our already established theories is part of what counts in the evidential evaluation of rival theoretical explanations for a given phenomenon. If it is added that those already established theories freely intertwine unobservables with observables—which many of them do—then Boyd can make the same point Kitcher does. I don't think that van Fraassen would find this Kitcher-Boyd argument unanswerable. The antirealist would likely point out in his own defense that this realist argument begs the question by assuming that success requires any sort of explanation in the first place. This would send the dispute back to the argument about cat-avoiding mice for another go-round. Even if van Fraassen granted the need to explain success, he would certainly deny that in our general scientific picture of the world, in our already established theories, there is no basis for distinguishing between observables and unobservables. This leads to the larger issue of whether the distinction can bare the epistemological weight antirealism requires it to bare.

Van Fraassen argues that science need only be empirically adequate, that its aim is only to produce theories all of whose *observational* sentences are true under at least one interpretation of them. This presupposes a clear and robust observational/theoretical distinction that packs a traditional epistemological punch. But we saw in chapter 3 how very questionable the observational/theoretical distinction is. It is an important aspect of this whole debate to note how much the antirealist needs such a distinction. The antirealist wants to disparage the nonapparent and highlight the apparent; hence, he needs a clear distinction between the apparent and the nonapparent, between the observable and the unobservable. If such a distinction cannot be made out, if the holists are correct and every claim in a theory possesses some nonnegligible degree of theoreticity (to use

Quine's word for it), then the antirealist will be thwarted in the very attempt to set up his view of things. Some realists have taken this to heart. The realist Grover Maxwell argued that realists ought to support the popular attempts since Kuhn's time to show that there is no clear observational/theoretical distinction, for if that can be done then the antirealist's task becomes nearly impossible to accomplish. If the observational is merely a lower grade of the theoretical, then van Fraassen's claim that science is required only to produce an account true to what is observed becomes rather hollow and without force. Therefore, van Fraassen must maintain that the internal structure of a science is not all of a piece, it is not all one web of beliefs—rather, it is demarcated into the observational and everything else.

Maxwell argued that any attempt to draw a hard and fast line between the observable and the unobservable was hopelessly arbitrary. In philosophy, a charge of arbitrariness is a terrible thing, for arbitrariness is taken to imply groundlessness, and groundlessness is taken to imply that the distinction in question is question-begging and rationally unjustifiable. Van Fraassen counters that as long as there are clear cases at the extremes, then the distinction is no more or less vague than any other distinction in our language. That is, as long as there are clear cases of observational terms and clear cases of theoretical terms, then the distinction is usable despite a large collection of in-between cases about which we are unclear (van Fraassen suggests that all our conceptual distinctions are vague and suffer from borderline cases that we are unclear how to classify). Van Fraassen takes Maxwell's challenge to amount to showing that there is a term that could *never* be used to refer to observable phenomena; for Maxwell had suggested that, under suitable conditions, virtually any term could be used to refer to observables. This is a bit odd, because one way of understanding the theory-ladenness of science is that it implies that every term in science is a theoretical one. Therefore, one would think that the real burden would be to show that there is a term which could *never* be used to refer to something theoretical. Nevertheless, van Fraassen rises to the converse challenge by arguing that a subatomic charged particle could never be observable; hence, the predicate expression '__is a charged particle' could never be an observational predicate. Van Fraassen imagines a defender of Maxwell's position countering to the effect that to a micron-sized intelligent being such a predicate might count as an observational one. Van Fraassen replies that this shows how "silly" the realist's position is, for observationality is defined relative to how *we* humans are *presently endowed* with receptor systems. The suggestion is that the realist is pulling off some kind of sharp practice by imagining micron-sized beings endowed with the same human receptor systems which we are endowed with. A being the size of the Milky Way galaxy might be able to observe the curvature of spacetime in some direct sense, van Fraassen imagines the realist arguing, therefore the curvature of spacetime is observable. Van Fraassen finds this argumentative move to be a "change of the subject," a form of the fallacy of equivocation committed by the realist. "Should I call the

Empire State Building *portable*" because an enormous giant could pick it up and carry it somewhere?, van Fraassen asks facetiously.

Van Fraassen makes a good point, but it is not quite the one he might think he makes. Observationality *is* defined relative to current human receptor systems; but, once we notice how constrained the ranges of those receptor systems are, it seems epistemologically perverse and rather chauvinistic to seek to restrict the knowledge claims of science solely to the stretches of nature within the unaided ranges of those receptor systems. There are so many other regions of nature to which our receptors do not have direct access. Why insist, as van Fraassen does insist, that science ought to have no business making claims about what goes on in those unobservable regions of nature? It is not as though the unobservable regions in question are not there—surely van Fraassen is not claiming that—so the attempt to restrict existence claims to only the observable seems a rather self-centered (human-centered) way of doing science. At this point in the debate van Fraassen certainly has more to say about the matter, but what he has to say amounts to canvassing the argument for realism we examined in section 10.2.1 and to declaring it to be bogus. We have already dealt with van Fraassen's complaints about the various versions of the convergence argument, so I will not repeat the details of them here. His basic conclusion is that no explanation of the general success of science is either appropriate or called for; therefore, he does not feel the push toward realism that derives from the massive and overwhelming convergence mentioned in the convergence argument for scientific realism. It seems as though we have a standoff of temperament. I say that because one side of the debate believes a certain state of affairs cries out for a certain explanation, and the other side just does not hear that state of affairs crying out at all. The realist says, "now just look at all that convergence!," and van Fraassen shrugs his shoulders in disinterest. Argument may have run out on us here and we are left drawing our mutually inconsistent lines in the sand.

10.3.3 Antirealism: Success Isn't What You Think It Is

There are antirealists who, unlike van Fraassen, acknowledge the massive convergence in science as requiring some sort of an explanation. These other antirealists deny, however, that realism is the appropriate explanation for the massive convergence in science. The convergence is due, rather, to the fact that the "text" of science is ultimately circular, that science cooks the data from the inside to get the positive convergence result; that is, the suggestion is that the convergence is not genuine convergence but an artificially produced phony convergence due to the fact that the conceptual schemes embedded in the theories used to observe nature already guarantee their own confirmations and justifications. The conceptual schemes presupposed by scientific theories and used to interpret experiments underwritten by those theories cut off practitioners from cognitive access to the sorts of experiences that would count as observing nonconvergent outcomes. Thus, the alleged massive convergence is really a kind of self-fulfilling and question-begging boosterism, nothing more than self-interested chest-thumping by science on its own behalf. This is the view of convergence one finds among the majority of social constructivists. From the previous two chapters it should be clear that Bloor, Latour, Woolgar, Shapin, Schaffer, Harding, and Longino do not find the convergence of experimental outcomes in science indicative of a practice-independent reality that our mature theories are getting mostly correct in the large. They each offer instead some kind of debunking explanation of convergence under which it is an artificial phenomenon mostly or wholly due to social and/or psychological factors that their studies claim to reveal. We spent the previous two chapters looking closely at these alternative debunking explanations of science's success, and we found them seriously defective. I will not repeat their defects here. What we should note at the present instead is an interesting contrast between van Fraassen and the social constructivists. Each supports antirealism, yet they differ in important ways on a number of epistemological issues. Van Fraassen is a believer in the observational/theoretical distinction, the social constructivists are not. The latter are, broadly speaking, holists, and van Fraassen has little fondness for holism. Van Fraassen does not denigrate the success of science by suggesting that it is an artificial phenomenon due largely to subjective projections onto nature by practitioners; the social constructivists by contrast do not respect the alleged success of science—for them it is not really objective success. What these differences show is how antirealism can cut across different epistemological categories, just like realism can cut across different epistemological categories (your author is one of but few holistic realists—most realists do not hold a holistic model of theories). The lesson is that we should not be hasty and presume someone must hold a particular ontological position solely because he or she holds a particular epistemological position. One must look carefully and closely at the total package, the total model of science being offered, to be able to tell which side of the metaphysical fence a philosopher of science is on.

10.3.4 Antirealism: Success May Have Nothing to Do with Truth

Finally we come to the arguments of Larry Laudan and Arthur Fine, both philosophers of science who are a bit difficult to place precisely on the map of philosophy of science. We will examine Laudan's position first. Laudan is no fan of social constructivism and relativism; yet, he is scathingly critical of the cosmic coincidence argument for scientific realism. Laudan argues that a theory's being true—even its being approximately true for the most part—is neither necessary nor sufficient for its being successful when applied to its appropriate domain of nature, if indeed it is successful in the first place. Laudan associates the cosmic coincidence argument with the claim that

if theory T's application meets with success, then T is true.

He allows the realist to weaken the consequent of this conditional to 'then T is approximately true', which many realists do. Laudan notes how often realists help themselves to a notion of approximate truth—the idea that a theory's errors are minor and mostly ones of detail while its major claims are correct—without first providing a means of measuring the amount of truth a theory is associated with. This is a genuine problem for realists, one first discussed by Popper under the rubric of verisimilitude (truth-likeness). Various attempts to present formal measurement procedures for ascertaining the truth-content of a theory, its degree of verisimilitude, have been made, none completely successful. To delve into this issue would require that we plunge into a degree of formal detail that is best left for another time and place. Most realists argue that, because realism is not dependent on any formal model of scientific theories under which their semantic properties must be precisely characterizable, the lack of a precise measurement procedure for verisimilitude does not destroy the realist position as a whole. Even on the weaker reading involving approximate truth, Laudan argues, the previous conditional is false as a matter of the history of science. Laudan claims that there have been many theories in the history of science that have been successful in the sense at issue in the cosmic coincidence argument but which were not true theories. In short, Laudan argues that there can be and have been many instances of false but successful theories. If he is correct about this historical point, so his argument runs, then truth cannot be the best explanation of why successful theories are successful.

Laudan's argument is heavily dependent on a particular historical interpretation of the history of science, one which finds that history full of lots of successful theories that were false and whose central theoretical terms failed to refer to anything at all. The realist would appear to have two basic avenues of rebuttal: either attack Laudan's reading of the history of science as being a misreading, or else grant that reading but argue that it doesn't refute the cosmic coincidence argument.

Philip Kitcher attacks Laudan along the second line of rebuttal. Kitcher argues that noting the mere falsity of a theory *as a whole* is not enough for Laudan to make his case. A theory as a whole can be false, yet large parts of it can be true. If the false parts of it have little or nothing to do with whatever level of success the use of that theory achieved, then the true parts of it can still be used to explain the success in question. This suggests that Laudan is guilty of a kind of division fallacy. What Laudan needs to do to make his case is to show how the false parts of the theory in question were essentially involved in the generation of whatever success was achieved through the use of the theory. Laudan does not do this: He rests content with pointing out the theory's falsity as a whole as though he assumes that the falsity spreads by division to each part of the theory. Kitcher suspects that, on a case by case basis, Laudan's historical illustrations will show that the false parts of the past theories in question had virtually nothing to do with whatever degree of success those theories were able to achieve.

I am dubious about the wisdom of talking about "parts" of theories in the way Kitcher does. Any philosopher of science sympathetic to Quinean holism will share my dubiousness. For a holist theories tend to be all of a piece, with the meanings of their constituent terms interconnected, and the assertability conditions of their constituent basic principles intermeshed. One implication of this interconnectedness is that falsehood tends to be "spread out" over a large chunk of theory rather than isolated in specific locations within the theory. By contrast, Kitcher's strategy for rebutting Laudan seems to presuppose that the false parts of a theory can be isolated in circumscribed pockets that leave the rest of the theory clean of the taint of error. Perhaps Kitcher can overcome these holistic worries of mine by providing a nonconventional method for decomposing theories into their proper parts, or perhaps he would simply argue that holism cannot be made consistent with realism. I hold the view that holism can be made consistent with realism. Accordingly, I choose the other strategy for rebutting Laudan. Laudan's historical case illustrations do not show what he claims they show. Laudan gives a list of theories from the history of science that he claims is composed of theories that were each highly successful yet false; hence, he argues, truth is not what accounts for success. The list of theories is rather surprising. It contains as member theories, for example, the vital force theories of physiology and the **humoral theory of medicine**.

I cannot imagine what success **vitalist biology** possibly could have met with. It was experimentally empty, to say the least. It produced no research tradition with any degree of achievement to it. What was daily practice like for a vitalistic biologist—building a vital-forcetometer? No, for there can be nothing to the daily practice of a vitalistic biologist: The vital force is nonphysical and therefore cannot be experimentally investigated as to its structure. The humoral theory of medicine was long-lived, but being long-lived is not the same as being successful. What diseases were cured by virtue of using the humoral theory of illness? Virtually none. In the course of nearly 2,000 years the humoral theory of medicine couldn't even figure out the real causes of infectious illnesses. In nearly 2,000 years of humoral medicine only very weak herbal antibiotics (of course humoralists didn't understand them as such) were developed—except for the element mercury, and its causal efficacy in treating infections remained a mystery in humoral theory. In nearly 2,000 years of humoral medicine the functions of various bodily organs were consistently misidentified, useless and dangerous therapies like bloodletting were foisted on patients, and no effective treatments for the vast majority of diseases were ever developed (King Louis XIII of France in the course of a single year underwent 47 bloodlettings, 212 enemas, and 215 purgings, all to remarkably little therapeutic effect). This is success? Perhaps Laudan confuses the metaphysical "explanations" and systematic but feckless "treatments" of humoralists with successful research. In their own words, of course, many humoralist physicians thought that they were doing great deeds—but who says we have to take their own words for the real story of what they were doing? Here we can see the value of Hacking's manipulability argument. Humoral medicine might have generated its own metaphysical explanations for things, and its own systematic but feckless treatments, but it did not produce the ability to manipulate successfully the posited unobservables it claimed existed. There were few, if any, successful interventions in the course of events under humoralist theories (in some cases there were no attempts to intervene at all, let alone failed attempts).

Clearly, Laudan does not mean by 'successful' what most realists mean by 'successful'. Just what he does mean is unclear. His list of successful but false theories contains an entry for theories of **spontaneous generation**. In this case, presumably, Laudan has in mind the situation in which ignorance helped our ancestors to misconstrue the real cause of an experimental outcome. Do you want to test the theory that maggots spontaneously generate inside of meat? Just leave a piece of meat out and watch. The flies buzzing around the meat, of course, don't seem relevant in any way under the theory, so we don't even bother to record their presence in the lab notebook. Eureka! A while later maggots appear in the meat. Our experiment was a success! But one experiment that goes according to plan does not amount to success in general. Try seeing if golden retriever puppies will spontaneously generate inside of meat—or how about lobsters, or clams? Throughout Laudan's argument there is a consistently loose use of the concepts of theory and success. A theory is virtually any general claim, seems to be his idea, and success is had if any sort of self-consistent speculative story, or any sort of systematic even if ineffective techniques, can be invented in which the alleged theory plays a key role. But mere systematic practice does not by itself amount to success. For example, astrologers have a systematic practice (there are rules for drawing birth charts, and so forth), and they *claim* success in the same way humoralist physicians claimed success in treating their patients for over a millennium; but no one except a new-age acolyte considers astrology to be a genuinely successful science.

The above three case illustrations of allegedly successful but false theories suggest that Laudan's list of such theories ought not to be taken at face value. It is not at all clear, and probably just plain false, that each of the three theories was successful in the sense at issue in the cosmic coincidence argument for scientific realism. Laudan has other theories on his list for which it may be more plausible to argue that they were false but successful in some minimal sense of successful. Even in those cases, however, the looseness of Laudan's notion of success ought to give us pause. Laudan's argument does serve a good purpose, however. It exposes the fact that most realists have long helped themselves to a concept of success that needs to be made more precise and more clear than it is in the current debate. Success, I have argued here, is more than mere systematic practice, more than merely telling consistent metaphysical stories about that systematic practice. It has something to do with successful manipulation under robust and independent conditions of variation, and with the ferreting out of microstructure. Precisely how much more it has to do with those indicators of reality is an ongoing project for realists to work on. At this stage of the debate, however, it is far too hasty to claim that Laudan's list proves that the realists are engaging in fictional history of science.

The philosopher of science Arthur Fine argues that we ought to distinguish between what he calls ground-level realism and a more global methodological realism. Within a particular domain of scientific practice practitioners often take success as evidential grounds for granting a kind of "approximate" truth to the theories used to produce the success. This is ground-level realism. Global methodological realism steps up a level to argue that the shared general methods by which the various domains of science achieve their separate successes are evidential grounds for granting a kind of approximate truth to the theoretical products of those shared methods. This is van Fraassen's ultimate argument for realism construed as a function of shared general methods.

Fine believes that Laudan's historical arguments, which we have just finished examining, show that ground-level realism is hopelessly question-begging. That is because Fine assumes realism presupposes a correspondence account of truth of a roughly Kantian sort. The realist, argues Fine, must mean by 'true' that the hypothesis in question matches or corresponds to some Kantian noumenal reality, some structured arrangement of things-in-themselves. We have encountered this argument before, of course. I argued in that previous encounter that any realist who admits that science seeks an account of Kantian noumenal things-in-themselves has given away the game, for noumenal things-in-themselves, by definition, are not accessible to epistemological contact through empirical experience. Hence, the foolish realist would have no basis on which to assert that mature scientific theories are true, because no correspondence relation could ever be confirmed to hold if one term of the correspondence is epistemologically inaccessible. The solution to the problem is the same one I recommended before: We must reject the Kantian metaphysics in which the phenomena/noumena distinction is characterized as a difference between possible and impossible epistemological accessibility.

Fine argues that global methodological realism is no better off than ground-level realism, for when all the polemical smoke has cleared it can be seen, he claims, that the justification for global realism ultimately relies on the justification for ground-level realism. Hence, if the latter begs the question through a failure to appreciate the significance of all the allegedly "successful" but false theories on Laudan's list, then surely the former does also. I couldn't agree more with Fine, for I'm not sure that his two types of realism amount to two distinct types at all. The global level of success is simply an inductive aftermath of the many ground-level successes. All the action will be at the ground-level anyway where individual cases of convergent success pile up into an overwhelming heap, which, if sufficiently overwhelming, then suggests that shared methods might be part of the explanation for all those previous separate successes in so many different domains. Because I find the Laudan-Fine argument that ground-level realism begs the question to be faulty (their historical case studies do not show that the history of science is filled to the brim with successful but false theories), I do not hold that global methodological realism begs the question on that ground either.

Fine claims to be just as critical of antirealism as he is of realism. Antirealism, he argues, tries to replace the realist concept of truth as a correspondence of scientific hypotheses to the structure of noumenal things-in-themselves with a concept of truth as some kind of contingent social consensus, for example, what is true is what would be agreed on by all rational inquirers in the limit of inquiry. But all such antirealist theories of truth, Fine claims, are simply dressed up forms of "behaviorism," attempts to analyze the meaning of a central philosophical concept, truth, solely in terms of the differences its use makes to human social intercourse. Behaviorism suffers from intractable problems of the same kind as realism, argues Fine; a behavioristic analysis of Peircean antirealism, for example, must be couched in terms of what rational inquirers *would* agree on in some forever fictional context called the limit of inquiry. But for someone to be in a position to ascertain what would be agreed on under such mythical circumstances is as epistemologically impossible as someone's being in a position to ascertain if a particular scientific hypothesis corresponds to a particular arrangement of things-in-themselves.

I do not think that antirealist concepts of truth are necessarily behavioristic in the sense Fine claims, but I shall not pursue the matter. It is of more interest to note that Fine believes we should adopt a philosophy of science that is neither realist nor antirealist. He calls this third alternative the **natural ontological attitude** (NOA). That is a rather annoying name for what I am confident the reader will shortly find to be a rather incongruous view of science. A number of philosophers of science find NOA to be simply inconsistent given what else Fine says about science. NOA requires, according to Fine, that we "accept" the results of science on their own terms as though they were on a par with "commonsense" knowledge. Well-confirmed scientific results are in no need of justification via some doctrinaire philosophical theory of what truth is. The results of science aren't so much true or false as they are trustworthy or untrustworthy. This sounds like a version of van Fraassen's argument that success in science needs no explanation at all, but it is in fact different. Van Fraassen grants that scientific theories over time have a slowly increasing truth content, that the success of science creeps ever higher on the scale; Fine holds that NOA grants no such progressiveness to science or its theories. Increasing success in the use of scientific theories, under the NOA, does not indicate that member theories in the historical sequence of scientific theories in question have increasing truth contents.

What is puzzling to realists is how much NOA seems like obviously inconsistent fence-sitting. Why should anyone accept the results of science as part of common sense if Fine's argument that success does not imply truth is correct? Why accept what is just as likely to be false as it is likely to be true? To put it another way, how can Fine square the claim that we should accept the results of science "as it is" with his arguments that nothing about science "as it is" suggests that its theories are true rather than false? Under those conditions, such acceptance appears less than rational. Worse yet, the massive convergent success of science simply goes *unexplained* under NOA, just as we saw that it does under van Fraassen's form of antirealism. One critic of Fine and NOA, Richard Schlagel, finds it utterly implausible that Fine could be serious when he argues that explanatory success could have *nothing at all* to do with truth. Could the massive success of contemporary genetics (alteration of genes, gene therapy, medical genetics, tissue typing in criminal trials, and so on) have taken place if it was *not* true that DNA has a double helix structure?, Schlagel asks incredulously. Schlagel finds equally far-fetched Fine's insistence that NOA is not committed to the thesis that science progresses over time, that later theories within a domain are more accurate than earlier theories. So, Schlagel argues, if Fine is right then we are not justified in being fairly sure that physics has progressed since the theories of Anaximander 2,500 years ago.

For my part the polemical tone in the words of both Fine and Schlagel illustrate my thesis that, when it comes to the realism/antirealism dispute, we often run out of argument and are reduced to differences in personal temperament. Fine and the antirealists are amazed that realists seem so lunkheaded as not to see how clearly they're begging the question. Realists are just as amazed that Fine and the antirealists seem so obtuse of mind as not to be embarrassed that they have no explanation whatsoever for what obviously cries out for explanation: the massive convergent success of the sciences. I once heard a philosopher exclaim with exasperated emotion, "But what other criterion for truth in science *could* there be except what works?" That philosopher didn't have it quite right. It should be, "But what other criterion of truth in science *could* there be but what works nearly always, works in many different contexts, tends to spread in its working to related domains, and works at a level of fine-grained (usually quantitative) detail that would positively daze the rest of the world if it knew of it?"

10.4 Conclusion

So, who is correct? Which side wins? I'm not sure that it matters nearly as much as the parties to the dispute might like to think it does, as long as it is not science itself that suffers harm in the course of the disputation. All sides to the realism/antirealism dispute admit that most quarters of science seem to progress quite nicely in blissful ignorance of the fact that there is a raging debate within philosophy of science about what science is, what it is doing, and where it is going. It is in the nature of powerful and established institutions to do this. After all, religion keeps on rolling along at full speed among the common people—the churches and mosques are built, the holy days are observed, and so on—despite several centuries of disputation by professional intellectuals over whether religion is all a bunch of fairy tales or not. The same goes for art, athletics, business, and other institutional big-name players in modern social life. This is not to suggest that the dispute is worthless. That the common folk, the workaday scientific technicians, are unaware that the Intellectual High-Muckety-Mucks are fighting it out about the nature of what they, the workaday technicians, are doing, certainly does not mean either that the nature of what they are doing is transparently obvious or that clarifying the nature of what they are doing would not be helpful. It means only that the practice of science has an inertia independent of our understanding of it, that it has ceased to be a philosophically self-conscious practice, and that it rolls on inexorably as a permanent fixture of the modern world. That it does roll on inexorably is, I believe, to our ultimate credit. I say this despite science's checkered past, despite its mistakes and travesties. For down the long haul of human time, no other social institution whose job it is or was to keep the epistemological part of the human project going has ever amassed such an impressive record of success—predictive success, explanatory success, and manipulative success. Ah, but see how my realist sympathies have returned the topic to success once again. I have never tried to hide from the reader my realist leanings. I can only hope that the experience of reading this work has provided the reader with a better appreciation for, with a knowledge of the arguments that can be marshalled for, the view that the burden of proof rests with those who would make science into merely one more failed attempt to find out—if I may be allowed a realist turn of phrase—the actual way things really are.

Further Readings

For C. S. Peirce's pragmatist characterizations of reality and truth in terms of the limit of inquiry see his, "How to Make Our Ideas Clear," in *Philosophical Writings of Peirce*, edited by J. Buchler, New York, Dover, 1955, pp. 23–41. Hilary Putnam discusses the cosmic coincidence argument, convergence in science, and introduces his distinction between internal realism and metaphysical realism in two classic papers "Reference and Understanding," and "Realism and Reason," both in his collection of papers *Meaning and the Moral Sciences*, London, Routledge and Kegan Paul, 1978, pp. 97–119 and 123–40 respectively. Ian Hacking's manipulability argument for scientific realism can be found in his *Representing and Intervening*, Cambridge, Cambridge University Press, 1983. John Searle provides a short and accessible critique of French deconstructionism as seen from an Anglo-American perspective in "The Word Turned Upside Down," *New York Review of Books 30* (Oct. 27, 1983): 74–79. Steve Fuller's suggested deconstruction of the realism/antirealism dispute is in his *Philosophy of Science and Its Discontents*, 2nd. ed., New York, Guilford, 1993. Bas van Fraassen takes on realism and defends antirealism in *The Scientific Image*, Oxford, Oxford University Press, 1980. He expands on his antirealism in *Laws and Symmetry*, Oxford, Oxford University Press, 1989. Larry Laudan attempts to refute the convergence argument in "A Confutation of Convergent Realism," *Philosophy of Science 48* (1981): 19–49, reprinted in *Scientific Realism*, edited by J. Leplin, Berkeley, University of California Press, 1984, pp. 218–249. The Leplin anthology contains a number of important papers on the subject of scientific realism/antirealism. Among them is Richard Boyd's defense of realism "The Current Status of Scientific Realism," pp. 41–82. Philip Kitcher rejects antirealism and defends realism in chapters 4 and 5 of *The*

Advancement of Science: Science without Legend, Objectivity without Illusion, Oxford, Oxford University Press 1993. Harry Collins's attempt to foist the experimenter's regress on realists is discussed in his *Changing Order: Replication and Induction in Scientific Practice*, Chicago, University of Chicago Press, 1992. Sylvia Culp deftly disposes of the alleged regress in "Objectivity in Experimental Inquiry: Breaking Data-Technique Circles," *Philosophy of Science 62* (1995): 438–58. Arthur Fine attempts to explain NOA in "The Natural Ontological Attitude," in *Scientific Realism*, edited by J. Leplin, Berkely, University of California Press, 1984, pp. 83–107, as well as in "And Not Anti-Realism Either," *Nous 18* (1984): 51–65. Richard Schlagel's critique of NOA can be found in "Fine's "Shaky Game" (And Why NOA Is No Ark for Science)," *Philosophy of Science 58* (1991): 307–23.

Glossary

Ad Hominem: Latin for 'to the person'. An erroneous mode of reasoning in which a personal attack on the holder of a belief is substituted for criticism of the holder's reasons for the belief.

Aleatory Model: A model of scientific explanation proposed by Paul Humphreys in which causes rather than laws of nature are what explain, where a cause is a factor that alters the probability of the event explained.

Anomalies: Something that is *anomos* (in ancient Greek) is something that is not consistent with law. In contemporary science anomalies are experimental results that are unanticipated and do not fit with the prevailing theoretical doctrines, models, or beliefs.

Antigen-Template Theory: A now discredited theory proposed in the 1930s to explain the formation of immune antibodies, which postulated that antigen, the substance eliciting the immune antibodies, acts as a kind of mold to shape the antibodies from generic precursor antibody molecules. Also called the Template Theory or the Instruction theory.

Antirealism: An antirealist holds that all facts of the matter are dependent on the conceptual frameworks (and the methods those frameworks underwrite) in which they are expressible or discoverable. To an antirealist, there can be no facts whose nature transcends our possible methods for discovering them, that is, there are no inquiry-independent realities in the limit of scientific inquiry. A *fictionalist* is an antirealist with respect to the existence of theoretical entities in science, that is, a fictionalist holds that theoretical entities do not really exist and theoretical terms are only pragmatically useful fictions. An *instrumentalist* is an antirealist with respect to mature scientific theories, that is, an instrumentalist holds that such theories are not literally true accounts of the phenomena they cover but only useful instruments for allowing us to organize our sensory experience. See **Realism**.

Antireductionism: A label for a host of related positions in the philosophy of science that hold that either no scientific theory is reducible to another or that one or more special-case scientific theories are not so reducible. See **Reduction**.

Bridge Laws: In certain formal models of reduction general regularities that relate terms unique to the reduced theory to terms in the reducing theory are called bridge laws. In Ernest Nagel's model of reduction bridge laws must be material biconditionals that relate coextensive predicates. See **Reduction** and **Material Biconditionals**.

Clonal Selection Theory: The currently held theory of how antibodies are formed in the immune system, which postulates that antibodies are made preformed by certain immune cells and antigen that happens to fit structurally a specific type of antibody interacts with that type so as to select it for further replication.

Convergence: A term for the situation in scientific practice in which different experimental procedures, methods, and manipulations independently produce outcomes that all presuppose the same theory or theories.

Correspondence-Rules: Also called c-rules. In the positivist model of science general postulates that define the meanings of theoretical terms exclusively in terms of observational terms are called correspondence-rules.

Cosmic Coincidence Argument: An argument for realism in science, which holds that when the convergence within a given domain of inquiry reaches a sufficiently high degree it would be absurd—the massive convergence would be a coincidence of implausibly cosmic proportions—to believe that the theories that underwrote the convergent experimental results were fundamentally false. See **Convergence** and **Realism**.

Deconstructionism: A recently influential method of mainly literary criticism, pioneered by a number of French intellectuals, in which the critic exposes the alleged incompleteness or incoherence of a given philosophical position using concepts and methods of argument presupposed by that philosophical position itself. To deconstruct a text is to show that it has no privileged "intended meaning," but rather many different meanings, and that there is no interpretation-independent fact of the matter that the text truly speaks about.

Deductive-Nomological Model: The most influential model of scientific explanation ever invented under which the explanation of an event requires that a description of it be deducible from the relevant laws of nature. Certain further conditions of a technical kind are imposed on the set of premises from which the deduction is made. Also called the Nomic Subsumption ("to bring under law") Model.

Dispositional Properties: Properties of a system which are such that their characteristic effects are manifested only under specific conditions, not under all circumstances. The system's ability to manifest the appropriate behavior under the specific conditions is a standing feature of its structure.

Epistemological: Epistemology is one of the five traditional subfields of philosophy. An epistemological issue is one that in some way has to do with knowledge: what knowledge is, how it is arrived at, its different sources, what forms it takes under what circumstances, how one justifies a claim to know something, how knowledge differs from accidentally true opinion, and so on.

Essence: A technical term in philosophical metaphysics. The essence of an entity consists of those properties of it necessary for it to retain its identity. Any property of a system which is such that if the system lost that property then it would cease to exist *as that system* is said to be an essential property of that system.

Experimenter's Regress: An argument championed by the social constructivist Harry Collins in which it is claimed that experimental outcomes cannot fairly be taken to confirm a test theory, because the same theory allegedly has to be used to calibrate the test device used in the experiment. If so, then taking the outcomes as confirming ones would be circular reasoning.

Falsificationism: A general philosophy of science proposed by Karl Popper in which the core of scientific method consists, not in trying to confirm hypotheses by subjecting them to experimental tests, but in trying to disconfirm them by experimental tests.

Feminist Science Criticism: A relatively new movement within contemporary philosophy of science that criticizes both standard science and standard philosophy of science from a feminist perspective, and which holds that gender-biases of a specially male kind infect mainstream science in ways that result in distortion and error.

Folk Psychology: A controversial term in contemporary philosophy of science that refers to the commonsense system of everyday beliefs any competent speaker of a language holds about the mental life of human beings. It is usual to represent folk psychology as an unrefined theory in which human behavior is explained by postulating internal mental states "inside" human persons called beliefs and desires whose contents are causally responsible for a good deal of their behavior. See **Propositional Attitudes**.

Forms: This term is the usual English translation of a word in ancient Greek first used by Plato to refer to an allegedly otherworldly realm of objective abstract entities whose natures fixed the natures of things in our own universe. For example, particular blue things are blue, Plato would have said, in virtue of "participating" in the Form of Blueness. Modern-day philosophers treat the Platonic theory of Forms with various degrees of humorous distaste.

Holism: A currently fashionable and increasingly popular philosophy of science (and philosophy of language) first proposed by W. V. O. Quine in which both the meanings of scientific terms and the disconfirmation of test hypotheses are holistic in nature. Technical terms are said to be introduced into science in large interdependent networks—not one at a time—and it is only large networks of hypotheses that meet the test of experimentation, not isolated claims. See **Underdetermination of Theory**.

Homeostatic Property Clusters: A phrase invented by the philosopher of science and scientific realist Richard Boyd, which refers to the occurrence in natural systems of clusters of properties that manifest a self-regulating stability under variance of environmental conditions. No member property is absolutely necessary to its given cluster, but the loss of any one member of a cluster results in an automatic adjustment elsewhere in the cluster that keeps the causal effects of the cluster constant.

Horror Autotoxicus: A now discredited law of early immunology which held that nature does not allow the evolution of organisms capable of launching an immunological attack against their own tissues or blood.

Humoral Theory of Medicine: The predominant theory of disease and its treatment in the Western World for nearly 2,000 years until the advent of causally effective medicine in the nineteenth century. Due largely to the speculative beliefs of a number of Greek physicians, the theory postulated that disease is a function of the imbalance of four basic bodily fluids called humors—blood, yellow bile, black bile, and phlegm—and that treatment called for reestablishing the proper balance through diet, bloodletting, purges, and herbal medicines.

Incommensurable: Literally, 'not comparable by a common measure'. Thomas Kuhn makes heavy use of the concept of two paradigms being incommensurable in his historicist model of science. A concept that is often misunderstood by nonphilosophers, the incommensurability of two paradigms would mean that we could not even say one was better than, more accurate than, or even explanatorily equivalent to, the other. See **Paradigm**.

Incrementalism: The term Thomas Kuhn used to describe the traditional belief that over the course of time truth slowly accumulates in science, that later theories absorb what was correct in earlier ones and add to it.

Internal Realism: A controversial form of realism proposed by Hilary Putnam in which ontological facts of the matter do exist but only relative to an overarching conceptual framework within which questions about them make sense and have a definite answer. Questions about the metaphysical legitimacy of the conceptual framework itself are rejected as unanswerable and based on the erroneous belief that we can come to know facts of the matter that are independent of any and all conceptual frameworks. See **Ontological** and **Metaphysical Realism**.

Interpetation: See **Model/Interpretation**.

Laws of Nature: One of the most loved and hated concepts in all of philosophy of science. Originally based on theological conceptions of the universe, the notion has since been cleansed of religious meaning. General regularities in natural systems that hold between different types of states of affairs are said to be laws of nature. For example, the general regularity, all NK cells secrete perforin when activated, relates lawfully the state of affairs of being an NK cell with the state of affairs of secreting perforin when activated. These two states of affairs are always so related, everywhere, and without exception. Laws of nature fix the causal structure of the universe and are ultimately what explain why what happens happened.

Limit of Scientific Inquiry: A concept of usefulness in metaphysics and philosophy of science first popularized by C. S. Peirce. Based on the mathematical concept of an asymptotic limit, the limit of scientific inquiry is a fictional time (that is, a time that we will never actually reach, for there will always be more time) in the future when the present methods of science have evolved to as close to perfection as they are capable of getting. Presumably, at the limit of scientific inquiry, there is no time left for science and knowledge to develop any further. Like a mathematical limit, we imagine approaching this fictional time of maximally mature science to an arbitrarily high degree of closeness without ever actually getting there.

Logical Positivism: A general philosophical movement of the twentieth century dating from the late 1920s and early 1930s in central Europe. The positivists made philosophy of science a central field of philosophy, and their model of science was sufficiently powerful and influential to have defined the philosophy of science for several subsequent generations.

Manipulability Argument: Ian Hacking's version of the cosmic coincidence argument for scientific realism in which the particular kind of convergent experimental success in question involves the manipulation of unobservable entities with complicated instruments. See **Convergence** and **Cosmic Coincidence Argument**.

Material Biconditionals: One of the five standard truth-functional logical connectives in mathematical logic is the material biconditional. Usually symbolized by the "triple bar" $\equiv$, the material biconditional represents the relation of one statement being necessary and sufficient for another statement. See **Necessary and Sufficient Conditions**.

Material Conditionals: One of the five standard truth-functional logical connectives in mathematical logic is the material conditional. Usually symbolized by the "horseshoe" $\supset$, the material conditional represents some but not all of the uses of the English 'if ..., then ...' form of speech, for its logical definition is technically problematic: A material conditional is false whenever its if-part is true and its then-part is false, and it is true in all other cases. This definition leads to certain odd technical consequences, which caused logicians to indicate its oddity by adding the adjective 'material' to its name to distinguish it from the logically strict conditional.

Mathematical Logic: The general system of logic worked out between 1870 and 1911 by Gottlob Frege, Bertrand Russell, and Alfred North Whitehead. Superior in certain ways to the preceding traditional logic that dated back to Aristotle, mathematical logic allowed the logical positivists to construct a powerful and fruitful formal model of scientific theories.

Metaphysical Realism: Hilary Putnam's term for the traditional kind of realism in philosophical metaphysics, which holds that there are objective facts of the matter completely independent of any and all conceptual frameworks in which those alleged facts might be expressible or discoverable. See **Internal Realism**.

Methodological: A general term for any issue involving the technical means of how practitioners in a domain of inquiry go about investigating and discovering the phenomena of that domain.

Model/Interpretation: In mathematical logic we must distinguish syntax from semantics. The symbols of a formal system are syntactic entities only (they are merely marks of certain shapes, sizes, and in a certain order); they have no intrinsic semantic meanings until we provide a formal semantic interpretation of the symbols. The technical term for such a formal semantic interpretation in mathematics is 'model.'

Naive Baconian: The English Renaissance figure Francis Bacon wrote an influential treatise on scientific method in which he argued that doing science was a matter of collecting experimental facts carefully in such a way that the one uniquely true theory would "announce itself" unambiguously in the collected factual data. This is now considered a horribly naive conception of science, and anyone who holds it is called a naive Baconian.

Naturalized Epistemology: A movement in contemporary epistemology that seeks to treat all epistemological issues as a posteriori ones. Naturalized epistemology treats our knowledge-gathering cognitive processes as processes within the natural universe, not as special "mental" capacities applied to experience from some a priori realm outside the material universe.

Natural Kinds: A central and controversial concept in metaphysics and philosophy of science that dates back to Aristotle. Roughly, a natural kind is a class of entities every member of which shares the same essential properties, and where the formation and continued existence of the class is a function of natural processes rather than of human social practices or conventions. NK cells and electrons, for example, are two different natural kinds. The contrast concept is that of an artificial kind, a kind that depends on changeable human social practices. Political moderates, for example, would be an artificial kind. If the universe contains natural kinds, then it comes with natural seams or joints, as it were, and in that case there would be facts of the matter independent of our conceptual frameworks of categorization. See **Essence**.

Naturalism: A term of broad meaning referring to a general philosophical tendency in which one attempts to understand phenomena as being wholly within the physical universe. The opposing view holds that certain phenomena are outside the physical world because they are abstract or mental or otherwise eternal and necessary phenomena. To "naturalize" a topic is to demystify it and make it something understandable in terms of physical processes.

Natural Ontological Attitude (NOA)**:** A position in the realism/antirealism dispute due to Arthur Fine in which one attempts both to accept the successful results of science as part of commonsense knowledge and to deny that the success represented by those results is evidence that the scientific theories that helped produce them are approximately true. Proponents of NOA commonly argue that scientific realism and scientific antirealism both beg the question, although in different ways.

Necessary and Sufficient Conditions: A statement X is necessary and sufficient for another statement Y whenever it is the case that X and Y are true in exactly the same situations and false in exactly the same situations. It is never the case that one is true when the other is false. See **Material Biconditionals**.

Nomological Danglers: The positivist Herbert Feigl's term for the laws of nature contained in a special science, a science that allegedly cannot be fit into the reductive hierarchy of scientific theories. If there were any special sciences in this sense, then their general laws would "dangle" outside the network of the rest of nature. See **Reduction** and **Unity of Science**.

Normal Science: Thomas Kuhn's term for the state of a science in which daily practice is controlled and managed by a prevailing paradigm. Normal science is inherently conservative, slow, and detailed. See **Paradigm**.

Noumena: A technical term in Kantian epistemology and metaphysics. Noumena are things-in-themselves, as contrasted with things-as-they-appear-to-us. Noumena are empirically unknowable somethings-or-others about which we can have no knowledge. See **Phenomena**.

Observational Terms: In the positivist model of science the terms of a scientific theory that refer to directly detectable objects, properties, or events are called observational terms.

Ockham's Razor: The name given to medieval philosopher William of Ockham's proposed general principle regarding explanatory adequacy: Never postulate more than the minimum number of different objects, properties, or events necessary to explain what one wants to explain. In other words, in doing science don't bloat the universe with superfluous entities.

Ontological: Ontology is one of the two traditional subfields of philosophical metaphysics (the other being cosmology, which in modern times has been taken over by astrophysics), *ont* being the present participle root of 'to be' in Ancient Greek. An ontological issue is one involving the very being of things, their existence, their possibility, necessity, or contingency, and so on, as contrasted for example with epistemological issues of how and when we could come to know or have evidence of their existence, possibility, necessity, or contingency. See **Epistemological**.

Operationalism: A version of the positivist model of scientific theories in which the correspondence-rules defining theoretical terms are all of a special kind: The observational side of the c-rule specifies a physical "operation" that must be performed by the practitioner to detect, measure, or manipulate the theoretical object or property defined in the c-rule. See **Correspondence-Rules**.

Paradigm: The central concept in Thomas Kuhn's model of science, and one that is infamously hard to find a precise definition of. A paradigm is, roughly, a scientific achievement (experimental, methodological, theoretical, or otherwise) so overwhelmingly impressive that it defines daily practice for a community of scientific practitioners in the sense of settling fundamental issues in that domain of inquiry and thereby providing the practitioners with specific research puzzles to be solved that have definite solutions.

The Pessimistic Induction: A philosophical position read off the history of science (in particular, the history of physics) in which it is claimed that any scientific theory, no matter how seemingly well confirmed, will be shown to be technically false within some sufficiently lengthy time, say, a couple of centuries.

Phenomena: A technical term in Kantian epistemology and metaphysics. Phenomena are things-as-they-appear-to-us, as contrasted with things-in-themselves. In the Kantian system only phenomena can be the subject of knowledge. See **Noumena**.

Positivist Model: The conceptually rich and highly influential model of science proposed by the logical positivists in which scientific theories are held to be formal first-order languages that contain theoretical laws of nature and correspondence-rules defining theoretical terms exclusively in terms of observational terms. Now discredited, the positivist model serves to define by opposition every other model of science currently of interest.

Pragmatism: A general movement in philosophy, of American origin, in which philosophical concepts and problems are interpreted with respect to their impact on and meaning for the everyday practice of living in the ordinary world. Pragmatists are generally hostile to any type of philosophy that mentalizes philosophical issues and thereby gives an overintellectualized treatment of them.

Proof: In mathematical logic proof is a purely syntactic notion, not a semantic one. A proof is any sequence of formal sentences in which the last one is derivable under the syntactic rules of inference from the others in the sequence. A proof of statement X from statements Y and Z in no way ensures that X is true, unless both Y and Z are true.

Propositional Attitudes: A philosophical term for those mental states whose identity is specified in terms of an attitude focused on the content of one or more propositions. Commonsense psychology involves attributing to humans various beliefs and desires that are understood as attitudes focused on the semantic content of propositions. For example, suppose Leslie believes that summer is pretty but winter is profound. What is the precise content of this mental state of Leslie's? It is that mental state involving an attitude of belief toward the proposition 'summer is pretty but winter is profound'. See **Folk Psychology**.

Pseudoscience: A term referring to any system of claims that aspires to be considered scientific but which, for various reasons of a methodological and/or epistemological nature, fails to meet the minimum conditions for being a science.

Quine-Duhem Thesis: A thesis of considerable controversy in contemporary philosophy of science, which holds that it is always possible, if one were willing to pay a high enough epistemological price, to render any test hypothesis consistent with any seemingly disconfirming experimental data. The Quine-Duhem Thesis is a generalization of the underdetermination of theory by observational data, and it is especially beloved by holists. See **Holism** and **Underdetermination of Theory**.

Realism: A central term in contemporary metaphysics and philosophy of science of such broad use as to be unhelpful unless teamed with an adjectival modifier. Roughly, a realist is anyone who holds that there are objective facts of the matter independent of the conceptual frameworks in which those facts are expressible or discoverable. A *scientific* realist holds such a view with respect to certain issues in science: for example, that the theoretical entities of a mature scientific theory really do exist and are pretty much like the theory says they are, or that mature scientific theories are true accounts of phenomena whose natures are at least partially independent of those therories.

Reduction: The fundamental relationship between scientific theories in the positivist model of science. To reduce one theory to another is to show by means of formal logical methods both that the fundamental laws of the reduced theory are special cases of the laws of the reducing theory, and that the objects, properties, and events presupposed by the reduced theory are really combinations of the objects, properties, and events spoken of in the reducing theory. See **Unity of Science**.

Reductionism: A methodological orientation within science and philosophy of science in which it is taken as established that reduction is a reasonable, desirable, and achievable goal in science and therefore ought to be pursued in research and welcomed if achieved. See **Reduction**.

Relativism: A general term for the view that what is true or a fact of the matter is so only relative to some overarching conceptual framework in which the truth or fact of the matter is expressible or discoverable.

Revolutionary Science: Thomas Kuhn's term for the state of a science when it has entered crisis and the prevailing paradigm is under open attack by some of the practitioners of the science. Fundamental issues in that domain are called into question and research is redirected from small puzzle solving to the fundamentals of the field.

Scientific Explanation: The type of explanation that occurs in mature scientific practice is taken to be an especially cognitively satisfying and pragmatically useful kind of explanation, hence philosophers of science assume that providing a clear and concise model of such explanation is one central goal of any philosophy of science. It is granted by all concerned that there are other kinds of explanation than scientific explanation, but the latter is held to be the most important kind of explanation for a host of epistemological and ontological reasons.

Scientific Revolution: Thomas Kuhn's term for a historical case of the successful overthrow of one scientific paradigm by another scientific paradigm that replaces it.

Side-Chain Model: Paul Ehrlich's model of how an antibody response is generated in which preformed antibodies resting on the surfaces of the immune cells that secrete them physically interact with antigen and are subsequently shed into the bloodstream in great numbers. While incorrect in detail, something basically like this is how an antibody response is produced.

Social Constructivism: A term referring to an increasingly popular philosophical position in which the objects of our knowledge are held to be either wholly or partly constituted by our coming to know them in the way we do. Basically, social constructivism is a variety of antirealism and relativism.

Spontaneous Generation: The hypothesis that a particular kind of living organism can, under specific conditions, spontaneously appear in either inorganic matter or in biologically unrelated organic matter.

Strong Programme: A school of social constructivist sociology of science in which it is a cardinal tenet that the truth of a belief is never an explanation for why it is held by those who hold it. The production of true beliefs is to be explained by reference to the same kinds of social factors that explain the production of false beliefs. See **Social Constructivism**.

Supervenience: A highly technical concept introduced into philosophy of science by Jaegwon Kim originally to save physicalism in philosophy of mind from antireductionist arguments, but which soon took on a philosophical life independent of that particular issue. It refers to the relation that obtains when one family of properties M depends determinatively on a different family of properties N without particular member properties of M being reducible to particular member properties of N. In such a case M supervenes on N. If M supervenes on N, then any two objects that have exactly the same N properties must have exactly the same M properties, but the converse does not have to hold (two objects can have the same M-properties when each has different N-properties).

Taxonomy: A taxonomy is a system of classificatory concepts used to sort the entities, properties, or events of a domain of inquiry into separate kinds of entities, properties, or events.

Theoretical Laws: Laws of nature that contain only theoretical terms are theoretical laws. See **Theoretical terms**.

Theoretical Terms: In the positivist model of science terms that refer to objects, properties, or events that are not directly detectable are said to be theoretical terms.

Theory: The central concept of analysis in philosophy of science since the time of Newton. Different philosophies of science present different accounts of what theories are, although all schools of thought agree that a theory is not a mental entity in the mind of a practitioner. See **Positivist Model**.

Underdetermination of Theory: A term referring to an epistemological impasse of sorts, in which all the possible observational evidence that could ever bear on a given domain of phenomena is compatible with more than one theory that explains that observational evidence. The different competing theories are mutually inconsistent with one another, so we cannot simply lump them together into one huge theory.

Unification: A term for a much sought-after goal in many areas of intellectual endeavor, including science: the explaining of many diverse phenomena using the same fundamental principles, regularities, and concepts. If distinct phenomena A, B, C, and D can each be explained or accounted for in terms of explanation E, then E is said to unify A, B, C, and D.

Unification Model: A model of scientific explanation proposed by Philip Kitcher in which the correct scientific explanation of a phenomenon is that explanation among all the competing explanations which is maximally unifying. See **Unification**.

Unity of Science: A methodologically and ontologically contentious doctrine of the logical positivists, which holds that all the individual sciences will eventually form an interconnected network, a reductive hierarchy, in which higher-level sciences are reducible to lower-level sciences all the way from astrophysical cosmology down to quantum physics and pure mathematics. The Unity of Science, if correct, entails that nature is "of one fabric," the differences between distinct phenomena being of degree, not kind, and that there are no special sciences whose laws would remain outside the reductive hierarchy.

Vitalist Biology: A now discredited school of biological doctrine in which living organisms are held to contain a nonphysical "vital force" that accounts for their being animate rather than inanimate. The vital force is generally alleged to be spiritual in nature, and therefore not amenable to investigation by physical apparatus or procedures of any kind.

Suggested Readings

General Texts. Peter Kosso presents a short survey of philosophy of science for beginners and nonspecialists in *Reading the Book of Nature*, Cambridge, Cambridge University Press, 1992. Anthony O'Hear provides a general introductory survey of philosophy of science from a specifically Popperian point of view in *An Introduction to the Philosophy of Science*, Oxford, Oxford University Press, 1989. Larry Laudan tries to breathe dramatic life into the relativism issue as it relates to philosophy of science by using a dialogue format in *Science and Relativism: Some Key Controversies in the Philosophy of Science*, Chicago, University of Chicago Press, 1990. Carl Hempel's *Philosophy of Natural Science*, Englewood Cliffs, New Jersey, Prentice-Hall, 1966 is well-known, short, to-the-point, but terribly dated. A highly readable historical treatment of philosophy of science can be found in John Losee's *A Historical Introduction to the Philosophy of Science*, 3rd. ed., Oxford, Oxford University Press, 1993. William Newton-Smith discusses the many epistemological and ontological implications of the Kuhnian model of science with accuracy and frankness in his underappreciated *The Rationality of Science*, London, Routledge and Kegan Paul, 1981. Although it is a specialized work, Alexander Rosenberg's *The Structure of Biological Science*, Cambridge, Cambridge University Press, 1985 contains an outstanding summary of general philosophy of science in the first two chapters. For another work in philosophy of science which uses examples and illustrations from immunology see Kenneth Schaffner's *Discovery and Explanation in Biology and Medicine*, Chicago, University of Chicago Press, 1993. Schaffner is a physician as well as a philosopher. At a more advanced level, Philip Kitcher reaches for a grand amalgamation of all the main orientations in current philosophy of science in his imposing but worthwhile *The Advancement of Science: Science without Legend, Objectivity without Illusion*, Oxford, Oxford University Press, 1993.

Anthologies. The new standard among anthologies of readings in philosophy of science is *The Philosophy of Science*, edited by R. Boyd, P. Gaspar. and J. Trout, Cambridge, Mass., MIT Press, 1991. This work is long, difficult, and pitched at a graduate student level in most cases, but it covers just about every aspect of philosophy of science. A more modest anthology which is reasonably accessible to undergraduates is *Scientific Knowledge: Basic Issues in the Philosophy of Science*, edited by J. Kourany, Belmont, Cal., Wadsworth, 1987. *Readings in the Philosophy of Science*, 2nd. ed., edited by B. Brody and R. Grandy, Englewood Cliffs, New Jersey, Prentice-Hall, 1989, is an old standby that is still around.

Index

References to illustrations are in boldface type